Learn By Doing: Trigonometry

An Active Approach to Learning Mathematics

Ryan Hobbs

© 2025 by Ryan Hobbs

All rights reserved. No part of this publication may be reproduced in any form without written permission from Delaware Spring Press. www.delawarespring.org

ISBN: 978-1-7332514-4-0

This book follows the structure of Abramson, Jay. *Algebra and Trigonometry*. OpenStax, 2015, which is available for free at https://openstax.org/details/books/algebra-and-trigonometry-2e.
The traditional lecture videos, linked by QR code at the end of each lesson, use the material from *Algebra and Trigonometry*, and are available for free at https://www.youtube.com/@rphobbs2002.

The Reason for this Workbook

I have been a college math lecturer for over ten years and it didn't take me long to develop my foremost principle of education—if students aren't awake then they aren't learning. In my opinion, for most math students, the traditional lecture approach has limited value. And so, I began a journey to develop a more interactive approach to the mathematics classroom. This workbook is the result.

However, there is much more here than just a series of worksheets, as I've also strived to incorporate the other principles of math education which I've discovered over the years.

Math is a language. Good teaching must translate this language of numbers into English. Language learning also requires "comprehensible input." It is good to be challenged and stretched, but when teaching is beyond a student's level of understanding they become overwhelmed and shut down. As math progresses, it is valuable to show students how more advance concepts relate to simpler concepts which they already know and understand. And, when possible, it helps to make math visual.

Yet perhaps most importantly, good teaching comes beside students when they need help. This is why every activity has a QR code which links to a walk-through video of the active learning lesson.

In other words, good teaching is very much like good tutoring, and that is the fundamental idea which has guided the development of this course.

How to Use this Book

Each activity in this workbook has a QR code in the top righthand corner. This QR code will take you to a YouTube video where I work and explain every problem and concept in that lesson. It is my attempt to use technology to come along side of a student just as I would do in a real classroom.

So, as you work the activity, start and stop the video, as needed, for help. Or, skip through any part of the video which you don't need. The back of the workbook also contains a solution guide for checking your answers.

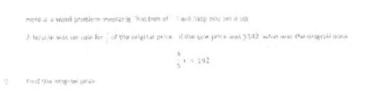

If after finishing an activity, you feel as if you could use some additional help, a QR code in the bottom right links to a traditional lecture for that section.

So, use the workbook as an instructional guide to a complete Trigonometry course, or use it as a tutoring resource to aid in a course you are already taking. This workbook is structured to follow the trigonometry portion of the textbook Algebra and Trigonometry. The activity numbers correspond to the chapters and sections of that book. For more practice problems, the textbook is available for free at https://openstax.org/details/books/algebra-and-trigonometry-2e.

Table of Contents

The Fundamentals of Trigonometry

- 7.1a: Two Systems for Measuring Angles and Converting Between Them 1
- 7.1b: Coterminal and Reference Angels 3
- 7.1c: Arc Length; Area of a Sector; Angular Speed 5
- 7.2a: The Six Basic Trig Functions 11
- 7.2b: Special Values of Trig Functions; Cofunctions 15
- 7.2c: Using Trig Functions to Solve Applied Problems 19
- 7.3a: Introduction to the Unit Circle 23
- 7.3b: The Pythagorean Identity; Sine, Cosine and Tangent of 30, 45, and 60 degrees 29
- 7.3c: Reference Angles; Completing the Unit Circle 33
- 7.4a: Finding the Exact Values of Tan, Csc, Sec, Cot 39
- 7.4b: Even/Odd Identities; Reciprocal Identities 41
- 7.4c: Alternative Form of the Pythagorean Identity; Finding Values of Trig Functions 45

Graphs of the Trig Functions and Inverse Trig Functions

- 8.1a: The Graph of the Sine and Cosine Function 49
- 8.1b: Transformations of the Sine and Cosine Function 57
- 8.2a: Graph of the Tangent Function 63
- 8.2b: Graphs of Csc and Sec 69
- 8.3a: Inverse Sine, Cosine and Tangent 73
- 8.3b: Composites of Trig Functions and their Inverses 77
- 8.3c: Inverse Functions with Algebraic Expressions 81

Trig Identities

- 9.1a: Verifying Trig Identities 83
- 9.1b: Using Even/Odd Identities; Using Algebra in Identities 87
- 9.2a: Using Sum or Difference Formulas 91
- 9.2b: Sum or Difference Formula for Tangent; Sum or Difference in Identities 95

9.3a: Double-Angle Formulas 97

9.3b: Power-Reducing Formulas 101

9.3c: Half-Angle Formulas 103

9.4a: Product to Sum Formulas 107

9.4b: Sum to Product Formulas 109

9.5a: Trig Equations with Sine and Cosine 111

9.5b: Trig Equations with Double Angles 117

9.5c: Trig Equations with Quadratics 121

The Law of Sines and Cosines; Using Trigonometry with Pre-Calculus

10.1a: Intro to the Law of Sines 125

10.1b: SSA and the Law of Sines 129

10.2a: The Law of Cosines 135

10.2b: Applied Problems with the Law of Cosines 139

10.3a: Polar Coordinates; Converting Between Systems 141

10.3b: Transforming Equations Between Rectangular and Polar 145

10.3c: Transforming Equations Between Polar and Rectangular 147

10.5a: Polar Form of Complex Numbers 151

10.5b: Products, Quotients and Powers of Complex Numbers in Polar Form 155

10.5c: Roots of Complex Numbers in Polar Form 159

10.8a: Introduction to Vectors 163

10.8b: Addition, Subtraction and Scalar Multiplication of Vectors; Unit Vectors 169

10.8c: Dot Product of Vectors; Angle Between Vectors 171

Trig

Active Learning: 7.1a

Below are three circles which each have the same angle. Below their picture, record the radius and arc length of each circle. (The arc length is the length of the 'crust' of the pizza.)

a)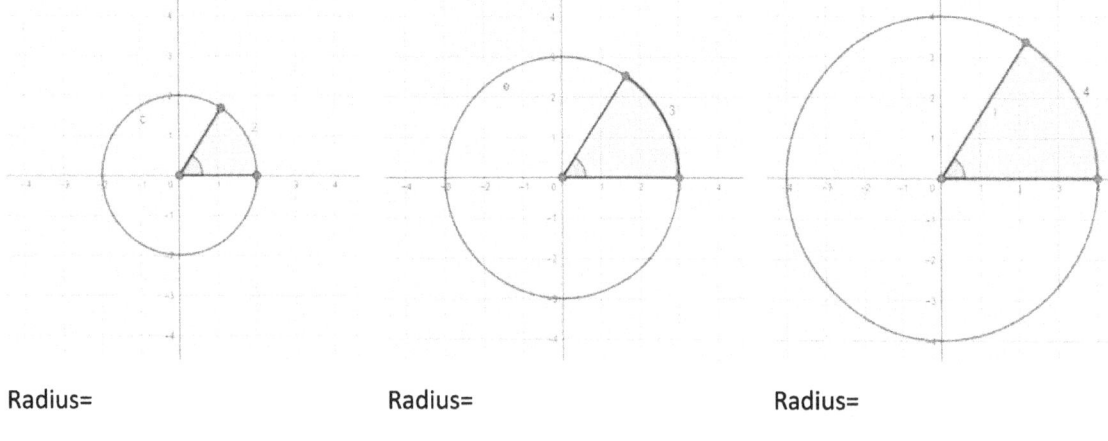

Radius=

Arc Length=

Radius=

Arc Length=

Radius=

Arc Length=

We are used to measuring angles in degrees, however there is another system for measuring angles called radians. The three circles above all have an angle of 1 radian. Finish the definition of a radian, by selecting the correct choice below. (Multiple choice.)

b) An angle of 1 radian, cuts an arc length which is equal to:

   a) The diameter of the circle.
   b) The radius of the circle.
   c) The circumference of the circle.

c) What is the radian measure of the angle below? (Hint: Think division.)

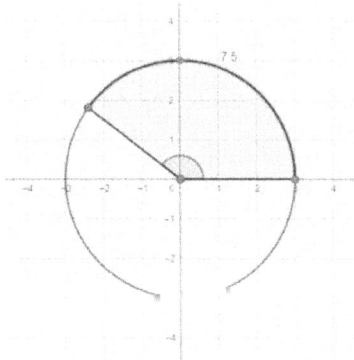

Most of the world uses the metric system, but the United States uses the standard system. Converting between units means to use the conversion ratio to cancel units.

$$1\ foot = .305\ meters$$

$$1\ kilometer\ (km) = 1,000\ meters$$

Here I'm converting 5 kilometers to feet:

$$5\ km \cdot \frac{1000\ m}{1\ km} \cdot \frac{1\ ft}{.305\ m} = \frac{5 \cdot 1000}{.305} = 16393.4\ feet$$

Notice that all the units cancelled and only the unit for feet remains so the final value is in feet.

d) Using the ratios above, **convert 25 kilometers to feet**. Use the approach I've just shown.

Converting between the two different systems of angles works the same way. There are 360 degrees in a circle and 2π radians. (Remember, π is 3.14.) So, π radians has 180 degrees.

$$\frac{\pi\ radians}{180\ degrees}$$

e) Convert 210° to radians.

f) Convert $\frac{\pi}{12}$ radians to degrees. (You will need to flip the conversion ratio upside down so the π cancels.)

Trig

Active Learning: 7.1b

Before we begin, we need to go over some basics of angles.

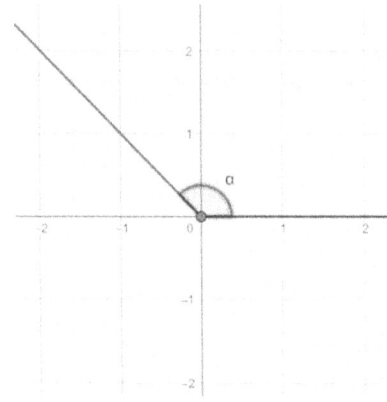

This angle is said to be in **standard position**. In standard position, we draw the angle on a coordinate plane. The starting side of the angle is called the **initial side**. In standard position, the initial side always goes along the x-axis. The ending side of the angle is called the **terminal side**.

In standard position, angles can be measured as either positive or negative. Positive angles are those which open counter-clockwise. Negative angles are those which open clockwise.

a) What do the following angles $30°$, $390°$, and $750°$ all have in common? Explain why.

Angles like these are called coterminal. Convert the angles below to a coterminal equivalent that is within one trip around the circle ($0°$ to $360°$).

b) 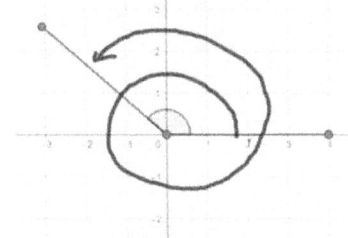 $500°$ (Think subtraction.)

3

c)   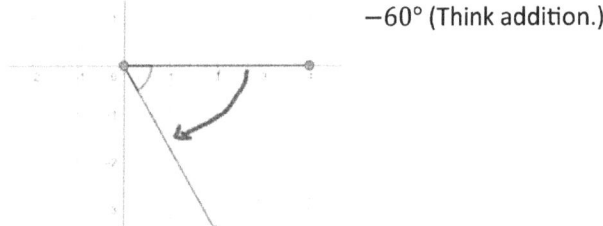   −60° (Think addition.)

d) We will do this again, but this time in radians. First, we need a trick. What is one trip around a circle when measured in radians?

e) Now write your answer as a fraction over 1.

f) What would be an equivalent measure of that fraction if the denominator was a 3?

g) What would be an equivalent measure of that fraction if the denominator was a 6?

Find a radian measure between 0 and 2π for the following angles. The trick is to add or subtract one trip around the circle, but use a version which has a common denominator with the angles below.

h) $\frac{8\pi}{3}$ (This angle is too large, so think subtraction.)

i) $-\frac{5\pi}{6}$ (This angle is too small, so think addition.)

Trig

Active Learning: 7.1c

a) Earlier we learned that radians are the arc length of a circle (the crust of the circle) divided by the radius. The formula is $\theta = \frac{s}{r}$. Find the radian measure of the angle in the picture below.

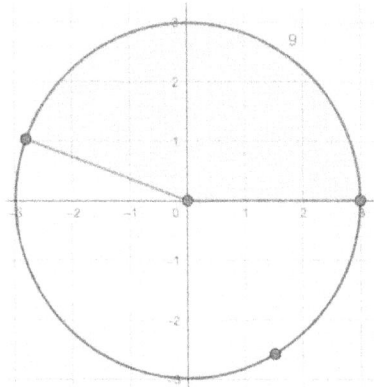

b) Suppose we wanted to find the arc length. Rearrange the formula to solve for arc length.

Find the arc length of the following:

c) 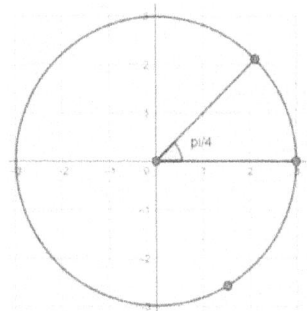 (The angle here is $\frac{\pi}{4}$)

d) 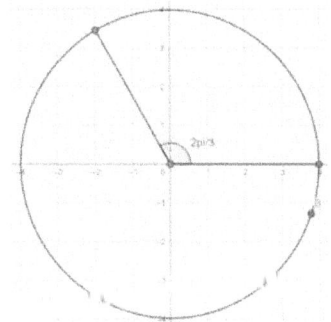 (The angle here is $\frac{2\pi}{3}$)

The formula for arc length *is* the definition of a radian, so when we work the problems, we need the angles to be in radians. Degree measures won't work. Here is one more problem, find the arc length. However, the angle is in degrees. First, convert the angle to radians.

e) 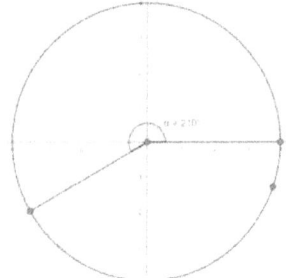 What is the measure of the angle in radians?

f) What is the arc length of the circle created by the angle?

A slice of pizza is exactly the same as our next topic. We want to find the area of a sector of a circle. A sector is like the pizza slice. There is a special formula to find the area of a sector but the idea is simple and the formula isn't necessary.

g) What is the formula for the area of a circle?

h) 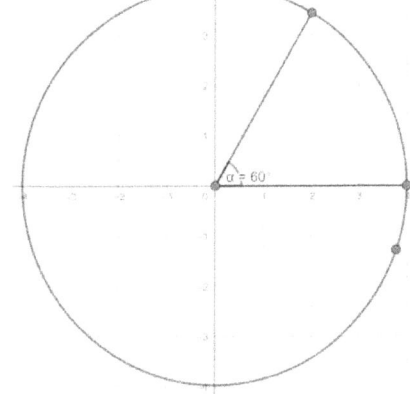 If a circle has 360° what percentage of the circle is the slice (the sector) in the picture below?

i) A sector is part of a circle. So, take the area of the circle and multiply it by the percentage. What is the area of this sector?

Try one in radians. The idea is the same.

j) 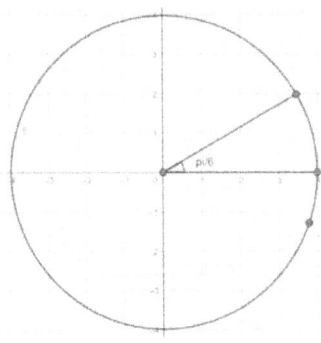 What percentage of the circle is this sector? (Remember we are in radians now, so the entire circle is 2π.)

k) What is the area of the sector?

Our next idea is that of angular speed (ω). Velocity is how fast something is going in a straight line. The formula is distance divided by time: $v = \frac{s}{t}$. But what if we wanted to know how fast someone ran through an angle? It is a strange idea to us because we don't usually think this way.

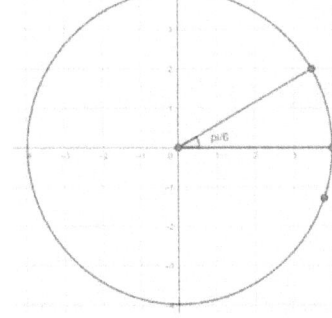

Suppose you are standing at the center of this circle watching someone run around the outside. Angular speed would be how fast they ran through the angle. The formula would be $\omega = \frac{\theta}{t}$.

l) What is the runner's angular speed if you watched them run through the angle $\frac{\pi}{6}$ in 20 seconds?

There is a simple formula that shows the relationship between velocity and angular speed: $v = r\omega$. The idea is easier to understand with an illustration.

Suppose you were again standing at the center of the circle. This time you are watching two people run. One runs on the inside circle and one runs on the outside circle. You say go and you watch them get to the end of a 45° angle in exactly the same time.

m) Are their angular speeds different or the same?

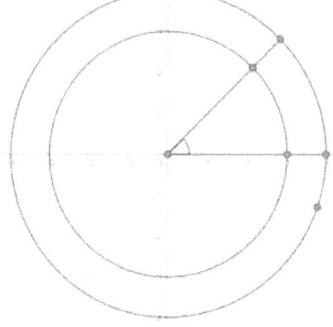

n) But which runner ran further, the runner on the inside circle or the runner on the outside circle?

o) So, which runner is running faster? (Their velocity is greater.)

Look again at the relationship between angular speed and velocity: $v = r\omega$. The r is the radius. Both runners had the same angular speed. The runner further out was running a circle with a larger radius. Therefore, when calculating their velocity, r would be a larger number, making their velocity greater. This makes intuitive sense. (Angular velocity, ω, needs radians. So, if you are given an angle in degrees, convert it.)

p) If a runner's angular speed is $5 \frac{radians}{minute}$ and they are running a circle with $radius = 10\ meters$, find their velocity.

q) If a runner is running a circle with $radius = 15\ meters$ and their velocity is $3 \frac{meters}{minute}$, find their angular speed.

Some problems may give angular velocity in revolutions per minute. But we need to be working in radians (not revolutions). A revolution is one trip around the circle, and one trip around a circle is 2π radians. Multiply the revolutions by 2π.

How many radians are in the following:

r) 3 revolutions

s) 100 revolutions

If we are given an angular speed in rotations per minute, we can simply convert the rotations to get rotations per minute. Here's an example:

Convert 120 rotations per minute into radians per minute.

$$120 \frac{rotations}{minute} \cdot \frac{2\pi \, radians}{1 \, rotation} = 240\pi \frac{radians}{minute}$$

Try one on your own.

t) Convert 300 rotations per minute into radians per minute.

Some problems will take this a step further.

A bike has wheels which are 30 inches in diameter. If the wheels are rotating at 120 rotations per minute, find the speed.

Speed is referring to velocity, and we have an angular speed in rotations per minute. We need the formula $v = r\omega$.

First, we find ω by converting the rotations per minute into radians per minute. (We did this above. It is $240\pi \frac{radians}{minute}$.)

Next, we need the radius of the bike wheel. The radius is half the diameter, so $r = 15 \, inches$.

Finally, we use the formula to find the velocity.

$$v = (15 \, inches)\left(240\pi \frac{radians}{minute}\right) = 3600\pi \frac{inches}{minute}$$

Try a similar problem on your own.

u) A car's wheels are 28 inches in diameter. If a wheel is rotating at 160 rotations per minute, find the speed.

Trig

Active Learning: 7.2a

Two right triangles are created below. Because they share the same angle, they are called similar triangles. Similar triangles are proportional. Find the length of the hypotenuse of the larger triangle by setting up a proportion and then solving. (Read the lengths of the known sides from the graph.)

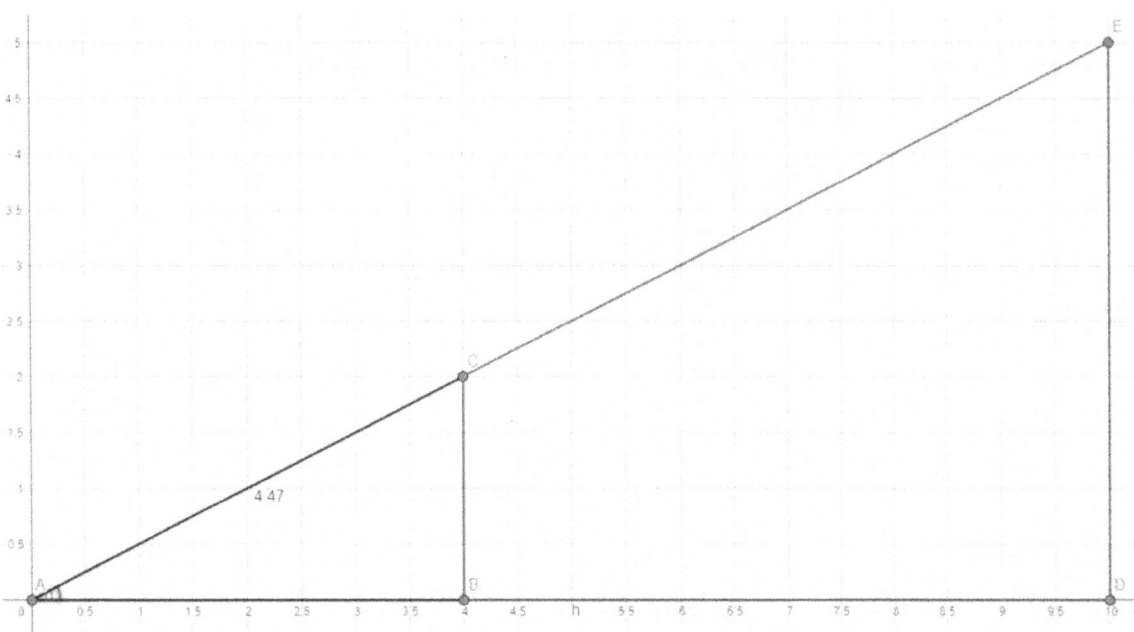

a) (Show your proportion here.)

b) Length of the hypotenuse $\overline{AE}$:

Trigonometry is based on a number of ratios. These ratios can be defined in terms of the sides of a triangle. Below are the definitions of the three most important ratios. They are in relation to the angle you are working with. (Opposite is across from the angle. Adjacent is next to the angle. And, hypotenuse is the hypotenuse of the triangle.)

Sine $\quad \sin t = \dfrac{opposite}{hypotenuse}$

Cosine $\quad \cos t = \dfrac{adjacent}{hypotenuse}$

Tangent $\quad \tan t = \dfrac{opposite}{adjacent}$

I want you to find the values of these trig ratios for both of the similar triangles.

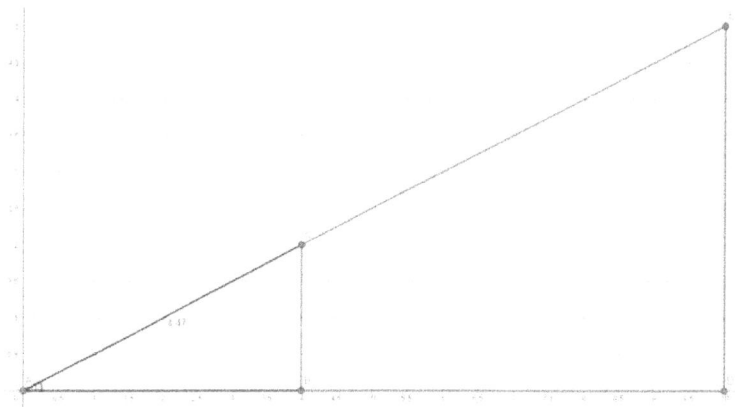

Smaller Triangle

c) $\sin \alpha =$

d) $\cos \alpha =$

e) $\tan \alpha =$

Larger Triangle

f) $\sin \alpha =$

g) $\cos \alpha =$

h) $\tan \alpha =$

i) What have you found to be true?

This is why the trig ratios are useful. Any right triangles built from an angle (as I did in the earlier exercise) are similar triangles. And in similar triangles any sine ratios will be the same, any cosine ratios will be the same, and any tangent ratios will be the same.

Using the triangle below, find the values of sine, cosine, and tangent for angle α.

j)

k)

l)

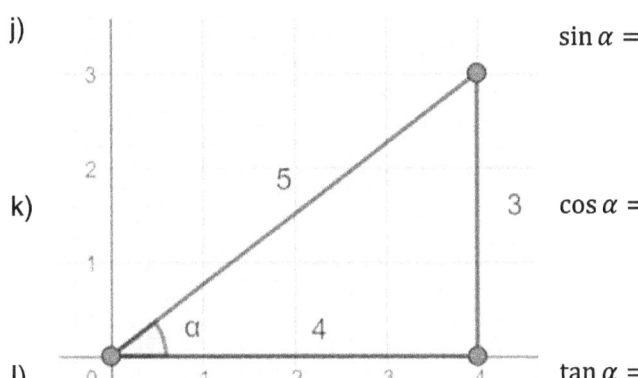

$\sin \alpha =$

$\cos \alpha =$

$\tan \alpha =$

There are three additional trig ratios that are important. They are the reciprocals (upside down) of the original three ratios:

Secant $\sec t = \dfrac{hypotenuse}{adjacent}$ Cosecant $\csc t = \dfrac{hypotenuse}{opposite}$ Cotangent $\cot t = \dfrac{adjacent}{opposite}$

$\sin t = \dfrac{1}{\csc t}$ $\csc t = \dfrac{1}{\sin t}$

$\cos t = \dfrac{1}{\sec t}$ $\sec t = \dfrac{1}{\cos t}$

$\tan t = \dfrac{1}{\cot t}$ $\cot t = \dfrac{1}{\tan t}$

Find the values of cosecant, secant, and tangent for the following triangle.

m)

n)

o)

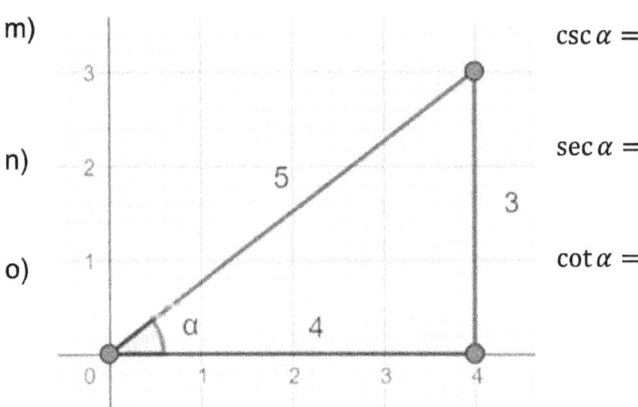

$\csc \alpha =$

$\sec \alpha =$

$\cot \alpha =$

Find the values of all six trigonometry ratios for this triangle.

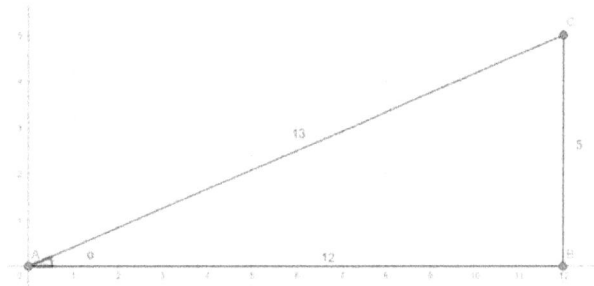

p) $\sin \alpha =$

q) $\csc \alpha =$

r) $\cos \alpha =$

s) $\sec \alpha =$

t) $\tan \alpha =$

u) $\cot \alpha =$

Trig

Active Lesson: 7.2b

Below is a 45-45-90 triangle. All such triangles are similar and have sides in ratio with the sides below. Use this triangle to find the value of the three basic trig functions for a 45° angle.

a)

b)

c)

$\sin 45° =$

$\cos 45° =$

$\tan 45° =$

The next triangle is a 30° − 60° − 90° triangle. These triangles are also similar and have sides in ratio with the sides below. Use this triangle to find the value of the three basic trig functions for 30° and 60° degree angles. (For the 60°, notice that you will be coming from the angle at the top right.)

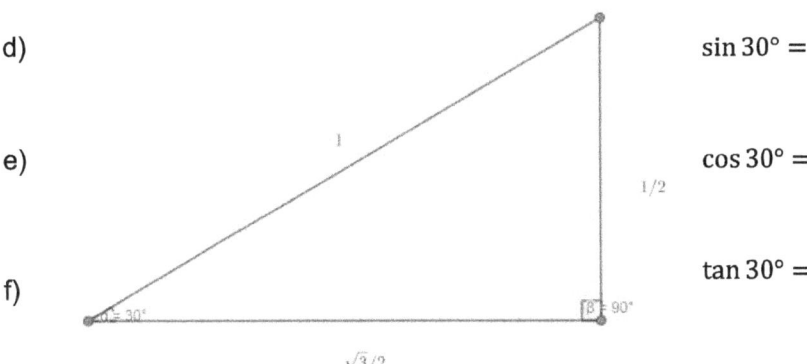

d)

e)

f)

$\sin 30° =$

$\cos 30° =$

$\tan 30° =$

g) $\sin 60° =$

h) $\cos 60° =$

i) $\tan 60° =$

j) In the 30° – 60° – 90° triangle, what do you notice about the value of sin 30° and cos 60°?

k) Look at the sides that these ratios are referring to and explain why this is happening.

l) What do you notice about the value of cos 30° and sin 60°?

m) Look at the sides that these ratios are referring to and explain why this is happening.

n) These are called cofunctions. There are others. In the triangle above, find the value of cot 60°.

o) What is the relationship between cot 60° and tan 30°?

p) Find the value of cot 30° and tan 60°.

q) Tangent and Cotangent are also cofunctions. Find the value of csc 30° and sec 60°.

r) What is true?

s) Find the value of csc 60° and sec 30°.

t) What is true?

Secant and cosecant are also cofunctions. All of the following are true. Look for a pattern regarding the angles.

$\sin 20° = \cos 70°$

$\cos 35° = \sin 55°$

$\sec 42° = \csc 48°$

$\csc 45° = \sec 45°$

$\tan 11° = \cot 79°$

$\cot 38° = \tan 52°$

u) What is true about the angles in cofunctions?

v) This is still true for cofunctions if they are measured in radians. Circle the one of the following which is *not* true:

a) $\sin \frac{\pi}{3} = \cos \frac{\pi}{6}$

b) $\tan \frac{\pi}{4} = \cot \frac{\pi}{3}$

c) $\csc \frac{\pi}{6} = \sec \frac{\pi}{3}$

Using what you know about cofunction identities, answer the following:

If $\csc 30° = 3$, find $\sec 60°$.

w) $\sec 60° =$

If $\sin \frac{\pi}{3} = \frac{\sqrt{3}}{2}$, find $\cos \frac{\pi}{6}$.

x) $\cos \frac{\pi}{6} =$

If $\tan 0° = 0$, find $\cot 90°$.

y) $\cot 90°$

Trig

Active Leearning: 7.2c

Suppose we wanted to find the length of the side $\overline{BC}$ in the triangle below. Since we know the angle and the adjacent side, we can set up an equation to solve for the missing side.

a) Only one of the following equations will work, circle the correct one:

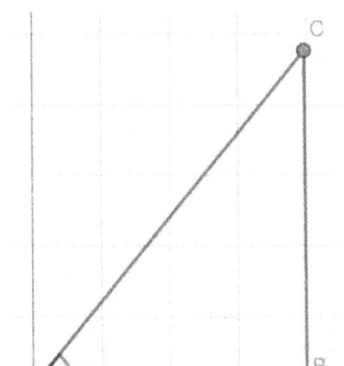

a) $\cos 50° = \dfrac{25}{\overline{BC}}$

b) $\tan 50° = \dfrac{\overline{BC}}{25}$

c) $\sin 50° = \dfrac{\overline{BC}}{25}$

b) The value of the trig function is simply a number you can find from your calculator. Find the value and solve for y. (Be careful: your calculator has both degree and radian settings. Make sure you are in degrees.)

c) Looking at the diagram below, which of the following would be the correct trig equation to solve for r

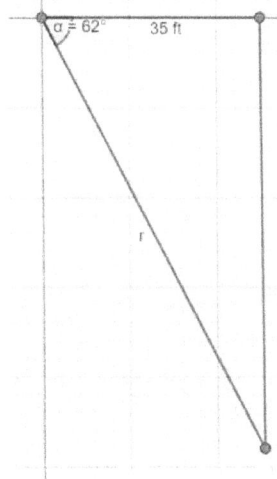

a) $\cos 62° = \dfrac{35}{r}$

b) $\tan 62° = \dfrac{35}{r}$

c) $\sin 62° = \dfrac{35}{r}$

d) Use your equation (and your calculator) to solve for r.

One common application of trigonometry is to find a missing distance. Before we work an applied problem, we need to understand some terminology. Imagine you were standing at the origin and looking forward. If you then looked up (creating an angle) you would create an angle of elevation. Instead, if you were looking forward and looked down, you would create an angle of depression. Below are the two graphs which we already worked with. One contains an angle of elevation and one contains an angle of depression. Put a circle around the angle of elevation. Put a square around the angle of depression.

e)

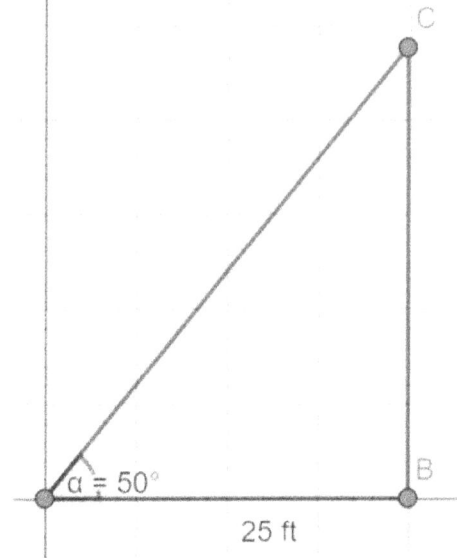

A 15-foot ladder rests against a wall and makes a 55° angle with the ground. The diagram has been created for you.

f)

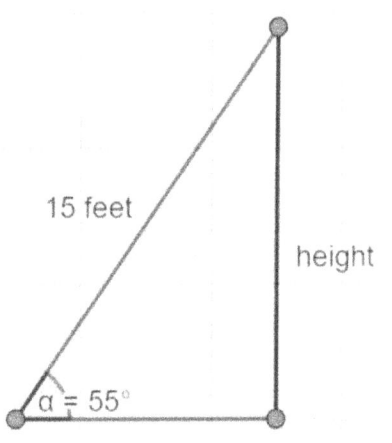

Find the height of the building at which the ladder is meeting the wall.

g) Is the 55° an angle of elevation or an angle of depression?

h) You are standing on a hill 22 feet away from the top of a lighthouse. Looking down at the bottom of the lighthouse creates an angle of $\frac{\pi}{4}$ radians. How tall is the lighthouse? (Either convert the angle to degrees or be sure to change the setting on your calculator to radians.)

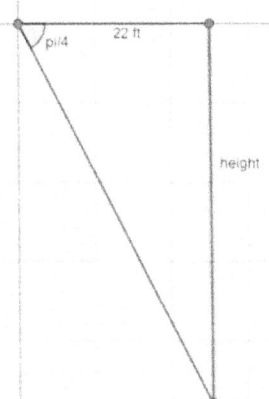

i) Is the angle, an angle of elevation or an angle of depression?

Trig

Active Learning: 7.3a

Below, I've given you the distance formula.

$$d = \sqrt{(x_2 - x_1)^2 + (y_2 - y_1)^2}$$

a)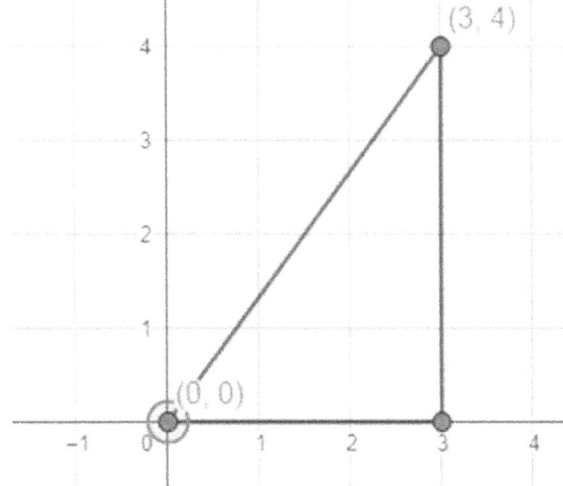

Use the formula to find the distance between the points $(0,0)$ and $(3,4)$.

b) There is a connection between the distance formula and the Pythagorean Theorem. What is it?

Using the triangle which I've made above, find the values of the basic trig functions. (α is the angle at the origin.)

c) $\sin \alpha =$

d) $\cos \alpha =$

e) $\tan \alpha =$

If we start a triangle at the origin, we can always think of the trig functions as:

$$\sin \alpha = \frac{y}{r}$$

$$\cos \alpha = \frac{x}{r}$$

$$\tan \alpha = \frac{y}{x}$$

Where r is the hypotenuse of the triangle.

In the picture below, find the values of the three basic trig functions. (First, you will need to use the Pythagorean Theorem to find the value of r.)

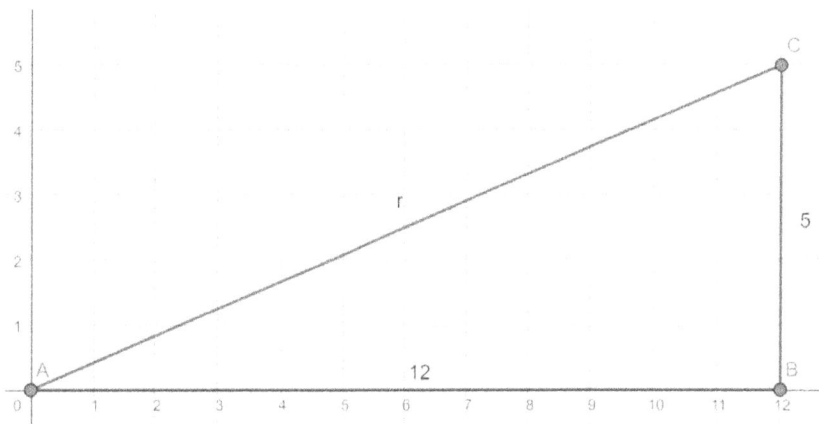

f) What is the value of r?

g) $\sin \alpha =$

h) $\cos \alpha =$

i) $\tan \alpha =$

The next trick is to put our triangles inside of something called the unit circle. The unit circle is the perfect circle with a radius of 1.

Here is an angle α within a unit circle:

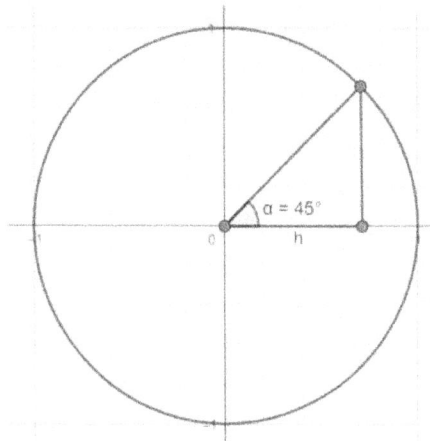

Inside of the unit circle r always equals 1. So, use this fact to simplify the following formulas?

j) $\sin \alpha = \dfrac{y}{r}$

k) $\cos \alpha = \dfrac{x}{r}$

I've added the coordinates of a point on the outside of the unit circle.

l)

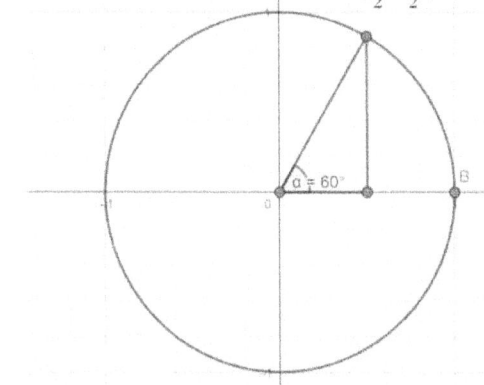

What are the values of x and y?

$x =$

$y =$

m)   In the unit circle, $x = \cos \alpha$ and $y = \sin \alpha$. So:

$\cos 60° =$

$\sin 60° =$

n)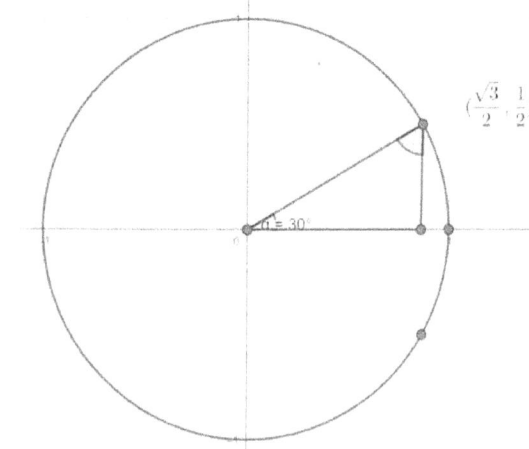

What are the values of x and y?

$x =$

$y =$

o) In the unit circle, $x = \cos \alpha$ and $y = \sin \alpha$. So:

$\cos 30° =$

$\sin 30° =$

p)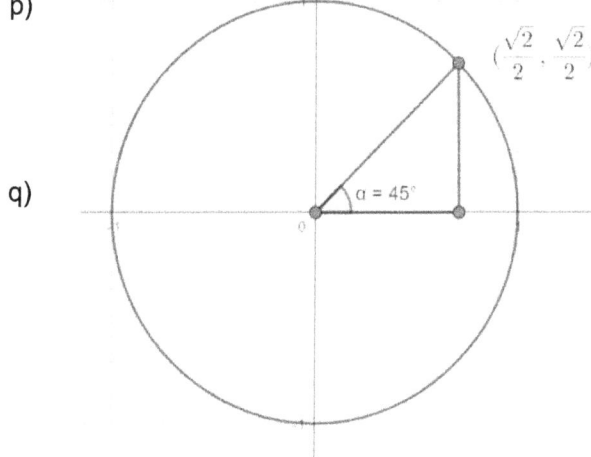

What are the values of x and y?

$x =$

$y =$

q) In the unit circle, $x = \cos \alpha$ and $y = \sin \alpha$. So:

$\cos 45° =$

$\sin 45° =$

Using the picture below, we will find the values of the three basic trig functions.

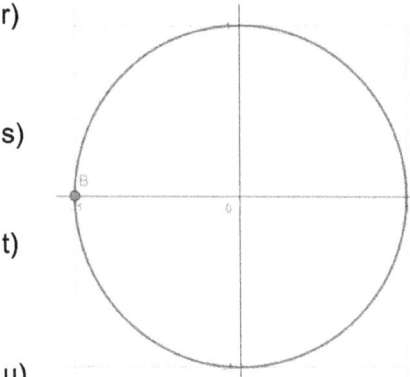

r) What are the coordinates of point B?

s) $\cos 180° = x =$

t) $\sin 180° = y =$

u) $\tan 180° = \dfrac{y}{x} =$

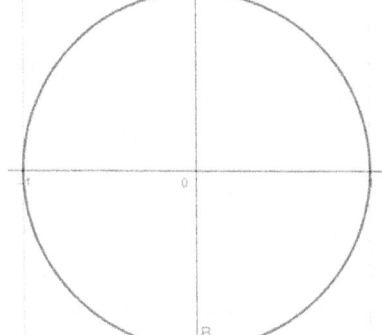

v) What are the coordinates of point B?

w) $\cos 270° = x =$

x) $\sin 270° = y =$

y) $\tan 270° = \dfrac{y}{x} =$

z) Notice that tangent encountered a mathematical problem. What was the problem?

aa) What would be the (x, y) coordinates for a point on the unit circle at π?

Find the value of the following:

bb) $\cos \pi = x =$

cc) $\sin \pi = y =$

dd) $\tan \pi = \dfrac{y}{x} =$

27

Trig

Active Learning: 7.3b

In our last activity, we learned that inside of a unit circle $y = \sin \alpha$ and $x = \cos \alpha$.

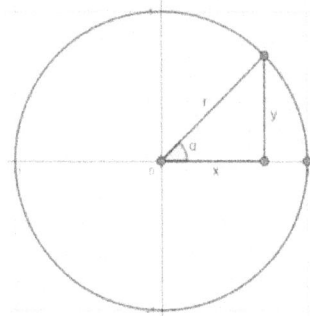

If $y = \sin \alpha$, what sign (positive or negative) does sine take in the following quadrants:

(Imagine a point in these quadrants. What sign does that y coordinate have?)

a)   Quadrant I

b)   Quadrant II

c)   Quadrant III

d)   Quadrant IV

If $x = \cos \alpha$, what sign would cosine take in the following quadrants:

e)   Quadrant I

f)   Quadrant II

g)   Quadrant III

h)   Quadrant IV

i) Using the labels from the graph below, state the Pythagorean theorem.

j) 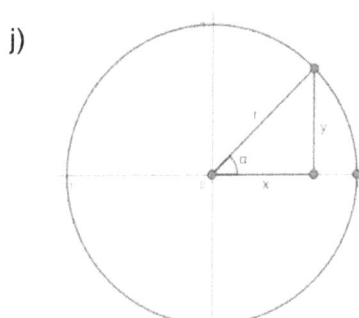 In a unit circle we know that $r = 1$, $y = \sin \alpha$, and $x = \cos \alpha$. In the space below, make these substitutions into the Pythagorean theorem.

In Trigonometry, the resulting equation is referred to as the Pythagorean Identity.

We will soon learn an easier way, but we can use the Pythagorean Identity to find a missing value of sine or cosine if the other value is known. If $\sin \alpha = \frac{3}{5}$ and α is in the second quadrant, find $\cos \alpha$. I've started the formula for you.

$$\left(\frac{3}{5}\right)^2 + \cos^2 = 1$$

k) Solve this equation for $\cos^2$.

l) Next take the square root of both sides. (Remember the square root of a number is really $\pm$).

m) The value of cosine isn't both positive and negative. We determine its sign by which quadrant the angle is in. We know the angle is in quadrant II. What is the sign of cosine in quadrant II?

n) Give the final value of cosine.

Try one on your own.

o) Use the Pythagorean Identity to find the value of sine if $\cos \alpha = \frac{5}{13}$ and α is in the fourth quadrant.

Trig

Active Learning: 7.3c

To finish understanding the value of the unit circle, we need an idea called reference angles. In the unit circle, we always start our angle along the positive side of the x-axis. That is called the initial side. The angle ends at the terminal side. A reference angle is the angle between the terminal side and the closest x-axis. And, it must always be positive.

a)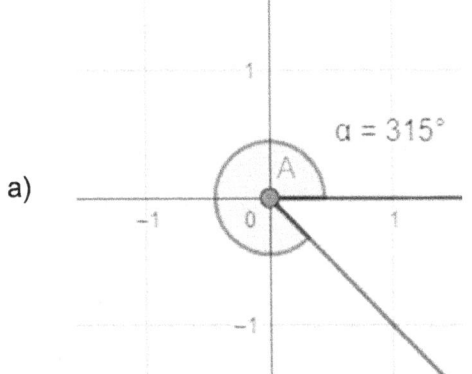

For this reference angle, the closest x-axis could be thought of as having an angle of 0° or 360°. Use 360° and subtract, so that we get a positive value. What is the reference angle?

$$360° - 315° =$$

In this next problem, the closest x-axis has an angle of 180°. Find the reference angle. (Again, the reference angle is always a positive angle. So, think of it as an absolute value or always subtract away from the larger angle.)

b)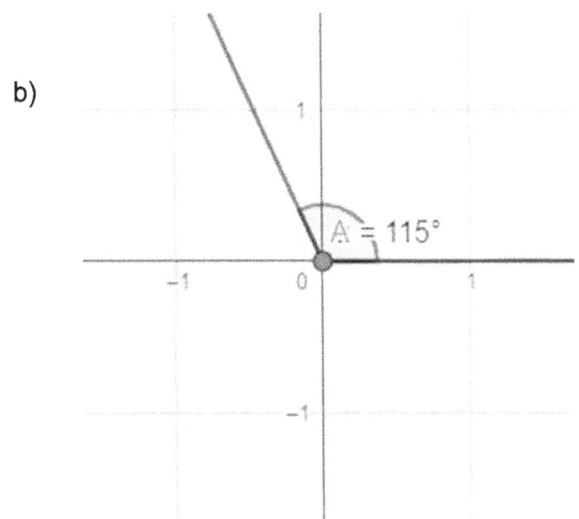

What is the reference angle?

33

The next problem is in radians and the closest x-axis would be π. Think of π as being $\frac{12\pi}{12}$. It is the same thing and the common denominator will make the subtraction easier.

c)

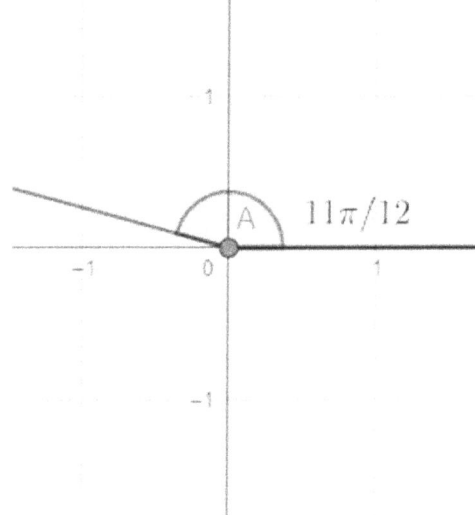

What is the reference angle?

Here, the closest x-axis would again be π. Think of π as $\frac{6\pi}{6}$.

d)

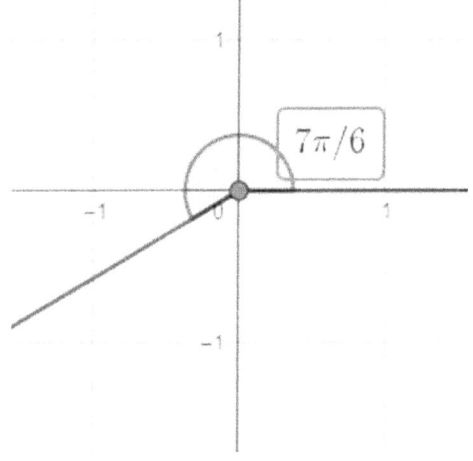

What is the reference angle?

Now, let's see why reference angles are important.

34

The graph below shows a 120° angle.

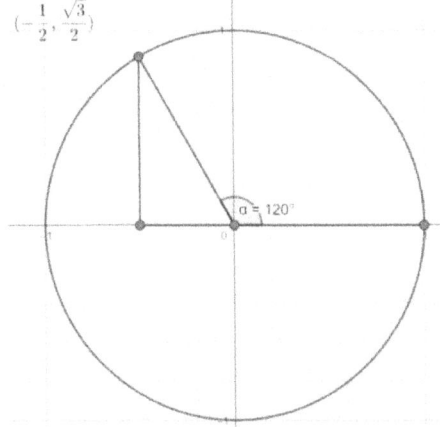

e) What is the reference angle?

f) I've given the coordinates of the point on the unit circle. What is the value of cosine?

g) What is the value of sine?

The values of cosine and sine are always identical to those values for their reference angle, except that they have been modified for the quadrant of the angle. Here, 120° is in the second quadrant.

h) Remember, cosine is x and sine is y. Is x positive or negative in quadrant II?

i) Is y positive or negative in quadrant II?

This is happening because the reference angles are making the identical triangles, just in a new quadrant.

This angle is 225°.

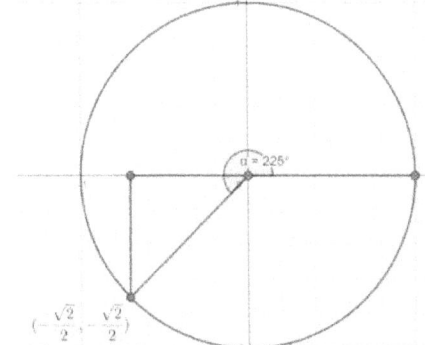

j) What is the reference angle?

k) Once again, I've given the coordinates of the point on the unit circle. What is the value of cosine?

l) What is the value of sine?

m) Again, these values are based on the reference angle, except that the signs are different. Why are the values of the cosine and sine negative for 225°? (Hint: think about the quadrant?)

n) With the idea of the reference angle, we can finish the unit circle. I've provided the first quadrant. (You have the angle in both degrees and radians.) Finish the other quadrants. (Remember, only the signs change.)

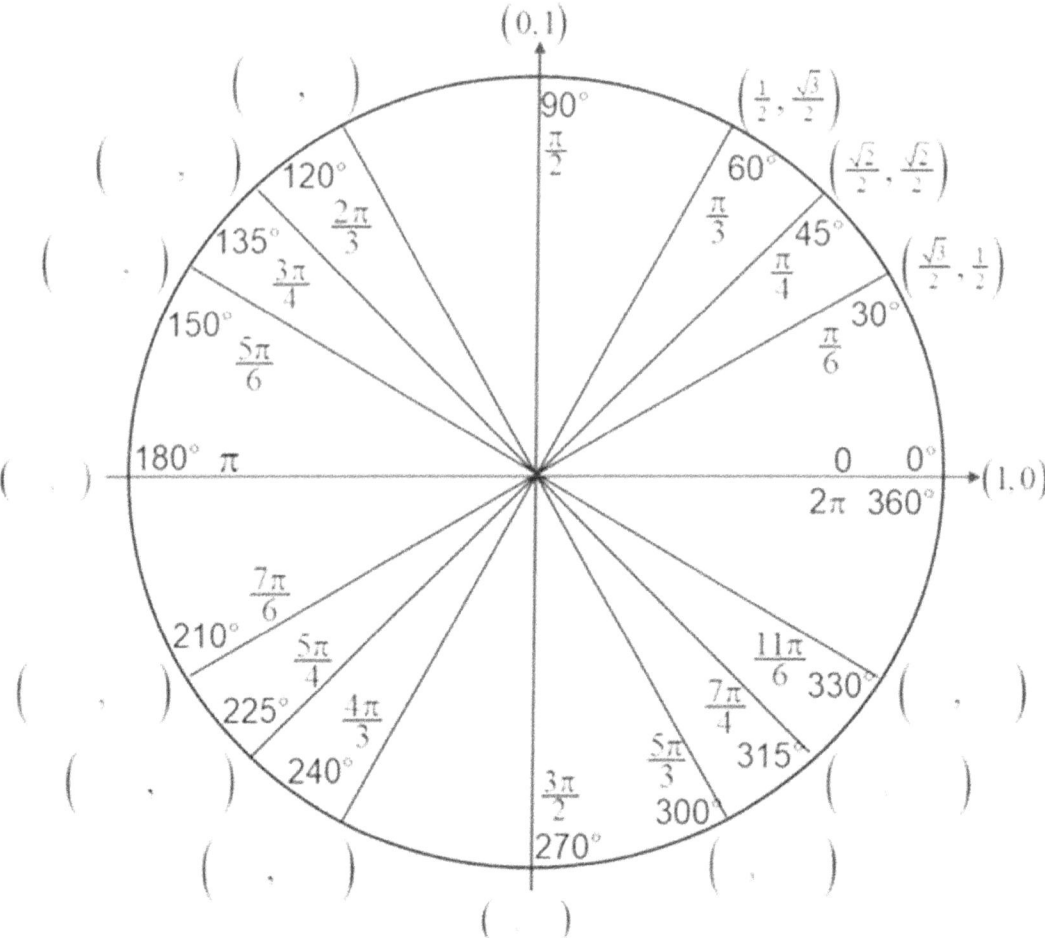

Here's an example.

150°

The reference angle is 30°. So, I will start with: $(\frac{\sqrt{3}}{2}, \frac{1}{2})$. However, cosine is based on x, and sine is based on y. In the second quadrant x is a negative and y is a positive. Therefore, my values for 150° are: $(-\frac{\sqrt{3}}{2}, \frac{1}{2})$

Use the same logic to complete the unit circle.

The unit circle is from Abramson, Jay. Algebra and Trigonometry. OpenStax, 2015, and is available for free at: https://openstax.org/books/algebra-and-trigonometry-2e/pages/7-3-unit-circle.

Once the unit circle is created, you can reference it to easily find values for the trig functions.

Use the unit circle to evaluate sine and cosine for each of the angles below.

- 120°

- $\dfrac{5\pi}{3}$

- 315°

- $\dfrac{11\pi}{6}$

Trig

Active Learning: 7.4a

If the unit circle gives us sine and cosine, we can easily compute tangent.

$$\tan \alpha = \frac{\sin \alpha}{\cos \alpha}$$

To find tangent, we only need to put the sine value over the cosine value. The only difficulty can be with the simplification. A fraction over a fraction is division. And, when we divide fractions, we Keep Change Flip. Keep the top fraction. Change to multiplication. Flip the bottom fraction. This first one is easy.

a) 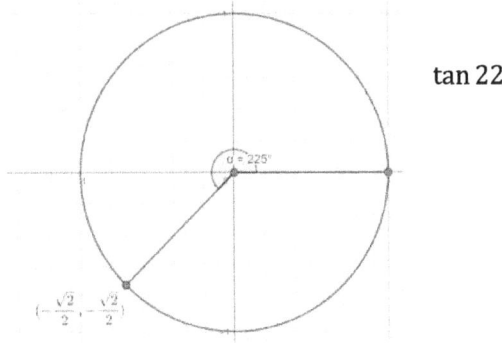 $\tan 225° =$

Try a harder one.

b) 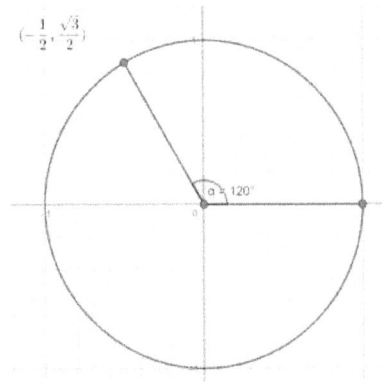 $\tan 120° =$

The process for cosecant, secant, and cotangent is even easier. Remember that the following are true:

$$\csc \alpha = \frac{1}{\sin \alpha}$$

$$\sec \alpha = \frac{1}{\cos \alpha}$$

$$\cot \alpha = \frac{1}{\tan \alpha}$$

So, simply turn the values for sine, cosine, and tangent upside down. (In trig, we won't leave square roots in the denominator, so you must rationalize the denominator.)

c) $\csc 225° =$

d) $\sec 225° =$

e) $\cot 225° =$

Try another:

f) $\csc 120° =$

g) $\sec 120° =$

h) $\cot 120° =$

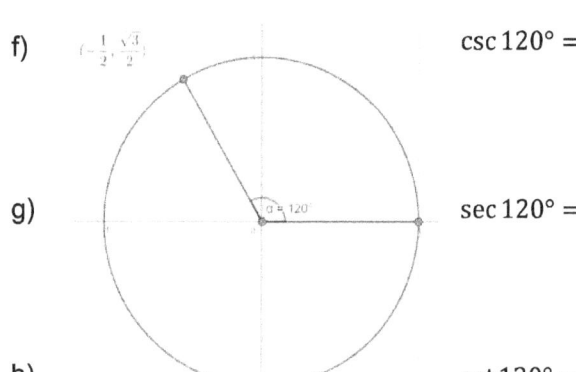

Trig

Active Learning: 7.4b

In Algebra II, we learn about even and odd functions. Here's a recap.

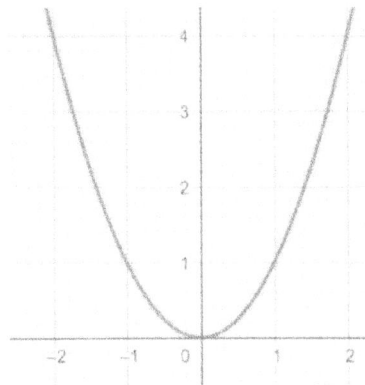

$f(x) = x^2$ is an even function.

Even functions are symmetrical over the y-axis. In other words, you could fold the graph over the y-axis and you would have an exact match on the other side.

You can test an even function by evaluating at $-x$. When you do, you get out the exact same function.

$$f(-x) = (-x)^2 = x^2$$

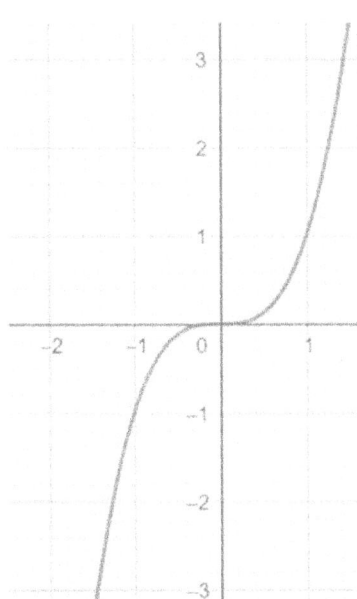

$f(x) = x^3$ is an odd function.

Odd functions are symmetrical over the origin. This means that you could fold the graph over the y-axis *and* then fold it again over the x-axis to make a match.

You can test an odd function by evaluating at $-x$. When you do, you get out the negative version of the original function.

$$f(-x) = (-x)^3 = -x^3 = -f(x)$$

As we will discuss more in a future section, trig ratios are functions too. And so, they also can be even or odd.

Here is the graph of the cosine function.

a) 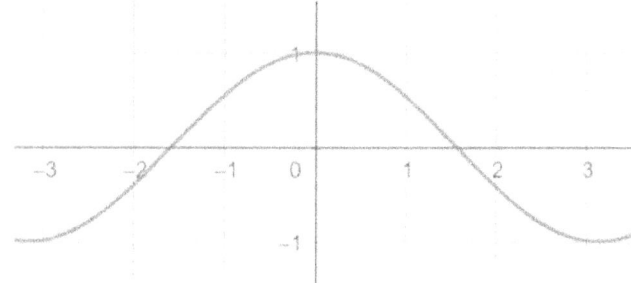 Based on the symmetry, is this graph even or odd?

b) So, which of the following is true:

    a) $\cos(-x) = \cos(x)$
    b) $\cos(-x) = -\cos(x)$

Here is the graph of the sine function.

c) 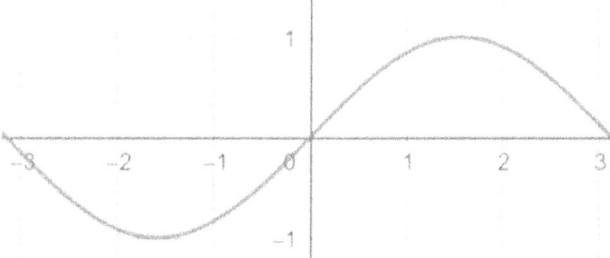 Based on the symmetry, is this graph even or odd?

d) So, which of the following is true:

    a) $\sin(-x) = \sin(x)$
    b) $\sin(-x) = -\sin(x)$

e) Which of the following trig functions is based on sine? (Circle the correct answer.)

$$\csc x$$

$$\sec x$$

f) Which of the following trig functions is based on cosine? (Circle the correct answer.)

$$\csc x$$

$$\sec x$$

These functions will follow the pattern of the trig function which they are based upon. With that in mind, finish the following:

g) $\csc(-x) =$

h) $\sec(-x) =$

Next, let's put $-x$ into tangent.

$$\tan(-x) = \frac{\sin(-x)}{\cos(-x)}$$

i) Which will be true:

   a) $\tan(-x) = \tan(x)$
   b) $\tan(-x) = -\tan(x)$

j) Is tangent an even or odd function?

Finally, let's put $-x$ into cotangent.

$$\cot(-x) = \frac{\cos(-x)}{\sin(-x)}$$

k) Which will be true:

   a) $\cot(-x) = \cot(x)$
   b) $\cot(-x) = -\cot(x)$

l) Is cotangent an even or odd function?

We could apply the idea in a problem like the one below.

If $\csc(x) = 2$, what does $\csc(-x)$ equal?

Since cosecant is an odd function $\csc(-x) = -\csc(x)$. So, if $\csc(x) = 2$ then the following must be true:

$$\csc(-x) = -\csc(x) = -2$$

Try these problems on your own.

m) If $\sec(x) = 2$, what does $\sec(-x)$ equal?

n) If $\cot(x) = 5$, what does $\cot(-x)$ equal?

43

Trig

Active Learning: 7.4c

Earlier, we learned how we could turn the Pythagorean theorem into the Pythagorean Identity.

$$cos^2(x) + sin^2(x) = 1$$

There are two other helpful versions of the Pythagorean Identity, and they are easy to create:

a)
- Divide each term in the Pythagorean Identity by $cos^2\alpha$, and simplify what you've made by using tangent and secant.

b)
- Now, divide the Pythagorean Identity again. This time, dividing each term by $sin^2\alpha$. Simplify what you've made by using cotangent and cosecant.

The unit circle below shows the value of cosine, but not of sine.

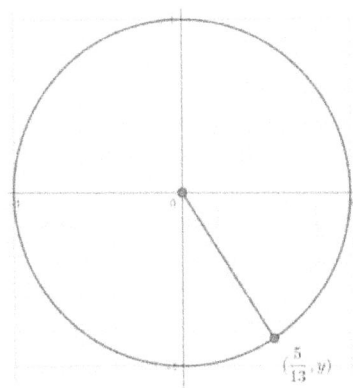

As we did earlier, we could use the Pythagorean Identity to find the value of sine, however there is an easier way.

We can simply create a triangle based upon what we know.

The triangle below has been built based on the value of cosine which was given to us:

c) 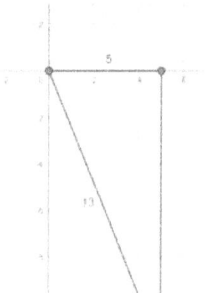 Use the Pythagorean Theorem to find the value of the missing side. What is the value?

d) This is quadrant IV, and the missing side is a value of y. Should y take on a positive or a negative value in this quadrant?

e) Our triangle is now complete and we can use it to find the value of sine. What is the value of sine?

Now that we have sine and cosine, find the values of the remaining four trig functions:

f) $\csc \alpha =$

g) $\sec \alpha =$

h) $\tan \alpha =$

i) $\cot \alpha =$

Try one on your own.

If $\sec \alpha = -\frac{5}{3}$ and is in quadrant III. Find the other five trig functions.

First, build the triangle. I've helped you.

j) 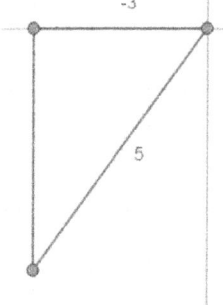 Use the Pythagorean theorem to find the missing side. Don't forget to assign a positive or negative sign based upon the quadrant.

46

Now that the triangle is complete. Use it to find the five trig functions.

k) $\sin \alpha =$

l) $\cos \alpha =$

m) $\tan \alpha =$

n) $\csc \alpha =$

o) $\sec \alpha =$

p) $\cot \alpha =$

Trig

Active Learning: 8.1a

a) In college algebra, the idea of a function was very important. We won't have a complete recap here, but a function could be identified by a vertical line test. If the vertical line (placed anywhere) would only hit the graph once, you had a function. Circle any of the following graphs which are functions.

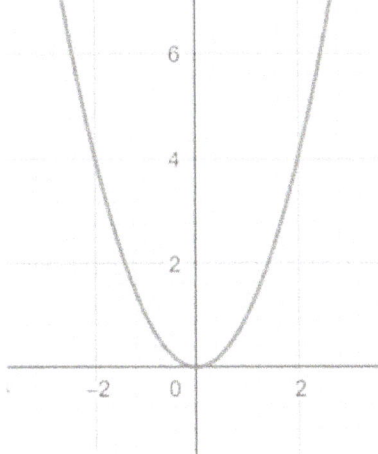
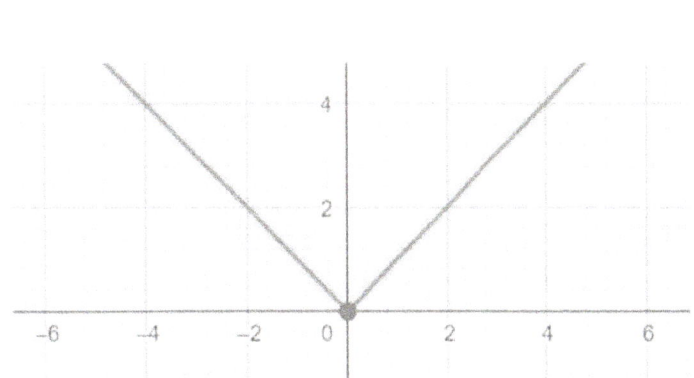

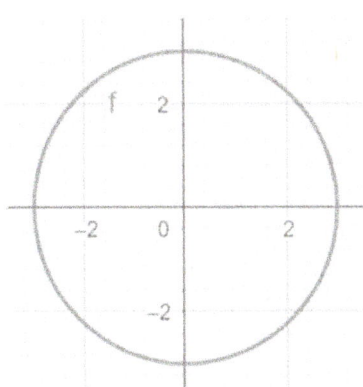

If you recall, $f(x)$ is just special notation for y. And when you have a function, you say y is a function of x.

The trig ratios can be turned into functions as well. They become functions of the angle. For instance, here is the sine function. $f(\theta) = \sin \theta$.

And, here's the graph:

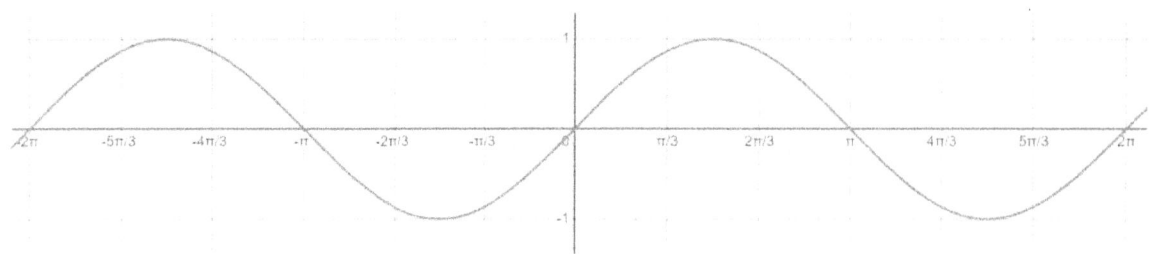

The angles are the input values and the ratios are the output values. Notice that this graph would pass the vertical line test. It makes a repeating wave because we are essentially stretching out the unit circle, which goes around and around. Now let's look at some important ideas.

b) Midline is the line which runs horizontally through the entire wave. Which of the following is the midline of this sine function:

                y-axis   x-axis

c) Amplitude is the distance from the midline to either the top of the wave or the bottom of the wave. It is always given as a positive distance. What is the amplitude of the sine wave?

d) Period is the length covered (on the x-axis) before the wave begins to repeats itself. Look at the graph of the sine function, what is the period?

Here is the graph of the cosine function: $f(\theta) = \cos\theta$

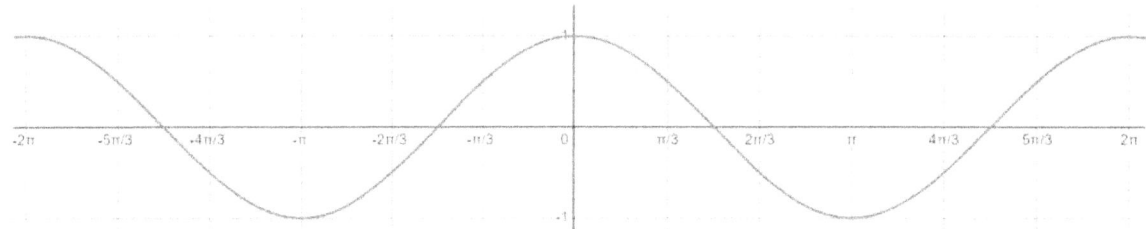

e) What is the midline?

f) What is the amplitude?

g) What is the period?

Here is the graph of the function $f(x) = 2\sin(x)$.

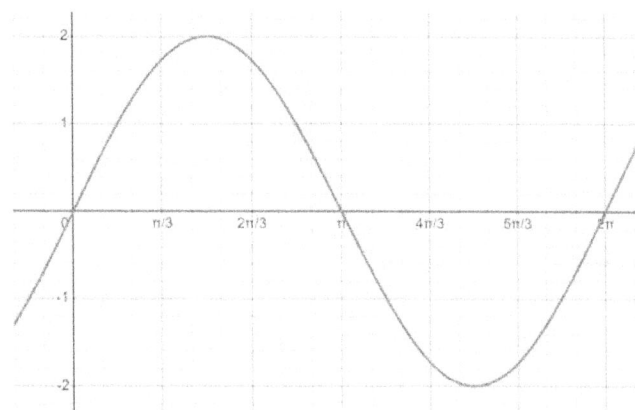

h) How has the function changed from the original?

i) Was this a change to the amplitude or the period?

When a change takes place on the outside of the function, it impacts the function vertically.

Here is the graph of the function $f(x) = -\frac{1}{2}\cos(x)$.

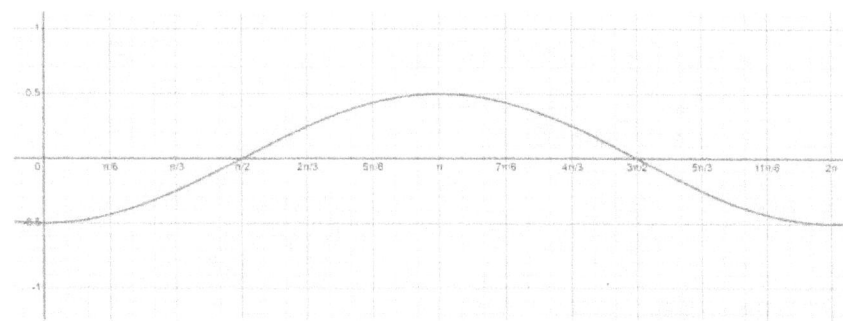

j) How has the function changed from the original? (Hint: There are two changes.)

51

Here is the graph of the function $f(x) = \sin(.5x)$.

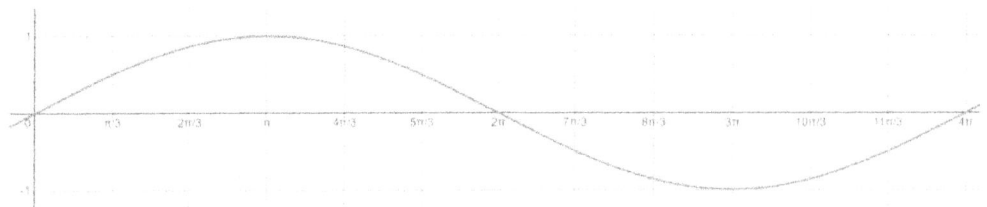

k) How has the function changed from the original?

l) Was this a change to the amplitude or the period?

When a change takes place on the inside of the function (with the x), it impacts the function horizontally.

Here is the graph of the function $f(x) = \cos(2x)$.

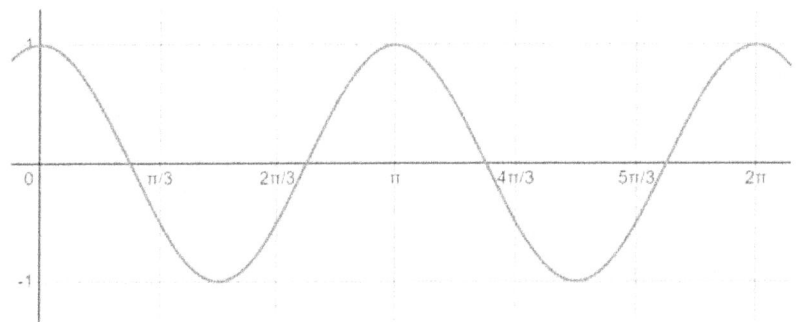

m) How has the function changed from the original?

n) Was this a change to the amplitude or the period?

Here is the graph of the function $f(x) = \sin(x) + 1$.

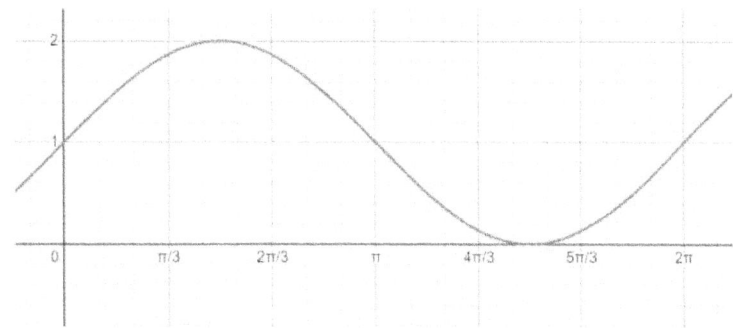

o) How has the function changed from the original?

p) The midline was the line $y = 0$. What is the equation for the new midline?

Here is the graph of the function $f(x) = \cos(x) - 1$.

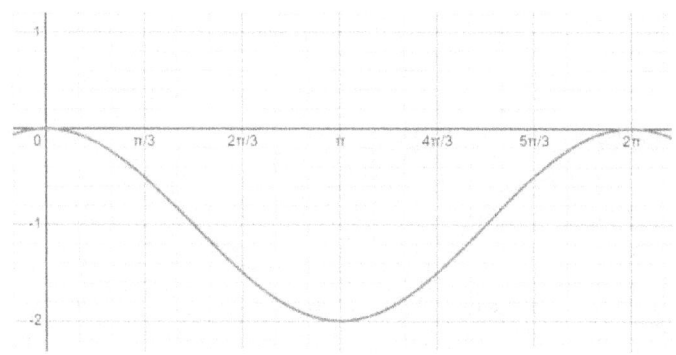

q) How has the function changed from the original?

r) The midline was the line $y = 0$. What is the equation for the new midline?

Here it the graph of the function $f(x) = \sin\left(x - \frac{\pi}{3}\right)$.

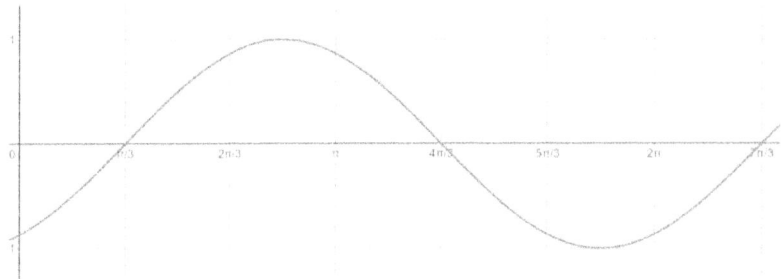

s) How has the function changed from the original?

Here it the graph of the function $f(x) = \cos\left(x + \frac{\pi}{3}\right)$.

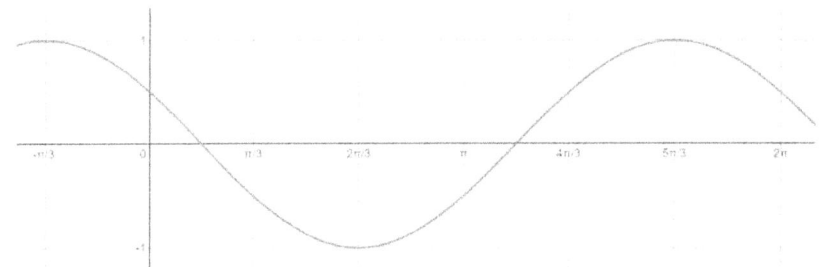

t) How has the function changed from the original?

These transformations are exactly the same as those we learned about in Algebra II. However, with trig functions they get special names.

- Horizontal shifts are referred to as phase shifts.
- Vertical stretches or compressions are changes to the amplitude.
- Horizontal stretches or compressions are changes to the period.
- Vertical shifts have moved the midline.

Remember, everything which is done vertically to a function is what you would expect. But anything done horizontally is the opposite of what you would expect.

Match the following:

u) $f(x) = 3 \sin x$ — A phase shift to the right $\frac{\pi}{2}$

v) $g(x) = \cos(x - \frac{\pi}{2})$ — Flipped over the x-axis

w) $h(x) = -\cos(x)$ — A vertical shift up 3. The midline is now $y = 3$.

x) $r(x) = \sin(2x)$ — An amplitude of 3.

y) $t(x) = \cos(x) + 3$ — A horizontal compression resulting in a period of π.

The period for both the cosine and sine functions is 2π. When a horizontal shift or compression occurs, the period is adjusted. To find the new period, divide 2π by the number in front of the x which is causing the horizontal shift or compression. For example:

$f(x) = \cos(4x)$

Period: $\frac{2\pi}{4} = \frac{\pi}{2}$

Find the period of these functions.

z) $f(x) = \sin(2x)$

Period:

aa) $g(x) = \cos(3x)$

Period:

bb) $h(x) = \sin\left(\frac{1}{2}x\right)$

Period:

Trig

Active Learning: 8.1b

In the last activity, we learned about the basic shape of the sine and cosine functions. Then, we learned how to transform (change) them. Next, we want to look at multiple transformations on the same function.

Look at this function:

$$(x) = 3\cos(2(x - \pi)) - 2$$

Amplitude: 3

Phase Shift: π to the right

Period: Cosine normally has a period of 2π but it has been horizontally compressed. To find the new period, take the old period and divide it by the number in front of the x.

$$\frac{2\pi}{2} = \pi$$

Midline: The graph has sifted down 2. The midline was the x-axis, which is the line $y = 0$. Moving the midline down, we now have the line $y = -2$.

Tell me the transformations that will take place to the following functions:

$$f(x) = .5 \sin\left(x - \frac{3\pi}{2}\right) + 1$$

a) Amplitude:

b) Phase Shift:

c) Period:

d) Midline:

e) Flips: Yes or No

$$f(x) = \sin(.5(x + \frac{\pi}{3}))$$

f)   Amplitude:

g)   Phase Shift:

h)   Period:

i)   Midline:

j)   Flips: Yes or No

$$f(x) = -10\cos(6(x + \frac{7\pi}{6}))$$

k)   Amplitude:

l)   Phase Shift:

m)   Period:

n)   Midline:

o)   Flips: Yes or No

p) Are the following two functions the same? If so, why? If not, why not?

$$f(x) = 10\cos(6(x + \frac{7\pi}{6})) \qquad f(x) = 10\cos(6x + 7\pi)$$

To ensure that you get the phase shift correct, take the number in front of the x and pull it out as a Greatest Common Factor (divide it out). Once factored, the phase shift is now isolated. It is simply the value added or subtracted to the x. In the example above, the phase shift is $\frac{7\pi}{6}$ to the left.

Find the phase shift of the following:

$$g(x) = 4\sin(2x - 2)$$

q) Phase Shift:

$$f(x) = -2\cos(3x + 6\pi)$$

r) Phase Shift:

$$f(x) = 3\sin(.5x + 2)$$

s) Phase Shift:

When graphing a transformed trig function, it works just as it did in Algebra II. Work your way from the inside out. (It must follow the order of operations.) First, do any phase shift. Then horizontal or vertical stretches or compressions. (Changes to the period or amplitude.) Any flips. And finally, any vertical shifts. (Changes to the midline.)

Look at this function:

$$f(x) = -2\cos\left(x + \frac{\pi}{2}\right)$$

t) Circle the correct graph:

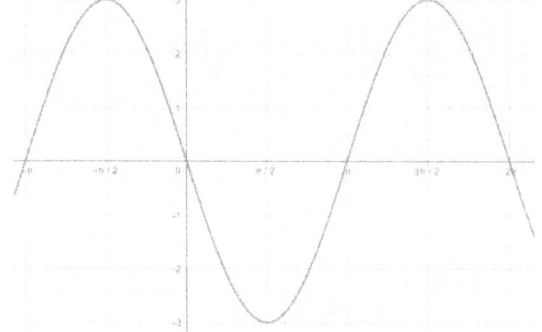 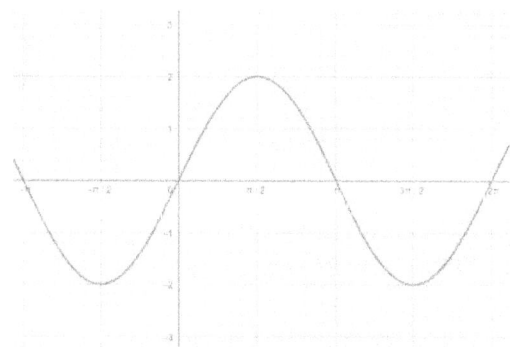

Try another:

$$g(x) = 3\sin\left(x - \frac{\pi}{3}\right) - 1$$

u) Circle the correct graph:

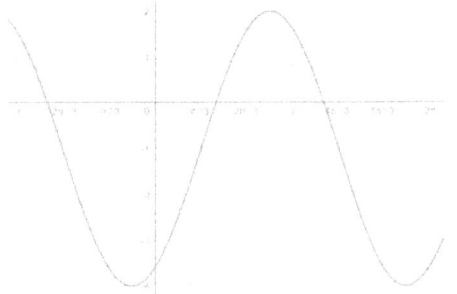 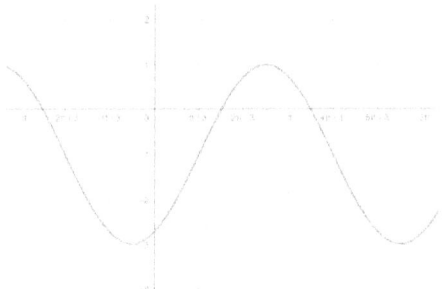

To find the value in front of x from a graph, we need to work the process in reverse.

$$Period = \frac{2\pi}{B}$$

(B is the number in front of x.)

Look at this graph. The period is the distance where the wave begins to repeat.

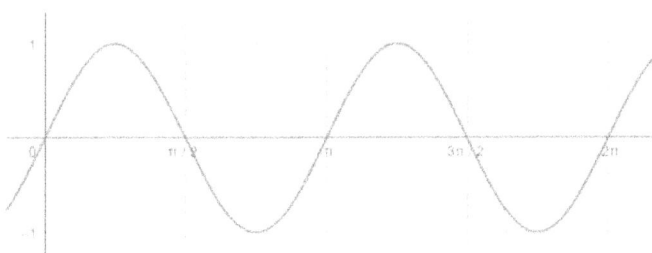

The period is π, so we have:

$$\pi = \frac{2\pi}{B}$$

Doing the algebra, we get:

$$B = \frac{2\pi}{\pi} = 2$$

Try some on your own.

Find the value of B.

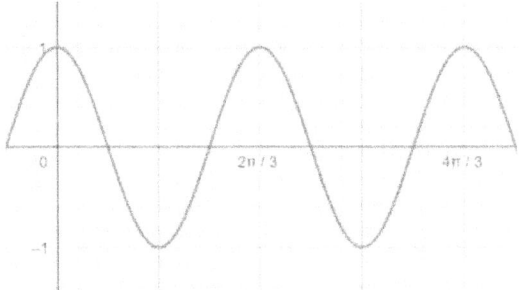

v) $B =$

Try another.

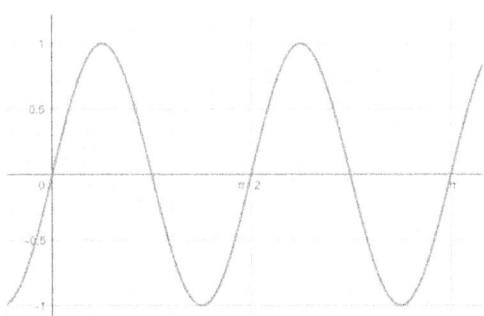

w) $B =$

Finally, let's put together an entire equation for a sine or cosine function based upon a graph. (I would not ask you to do this for an equation involving a phase shift.) Look at the graph below:

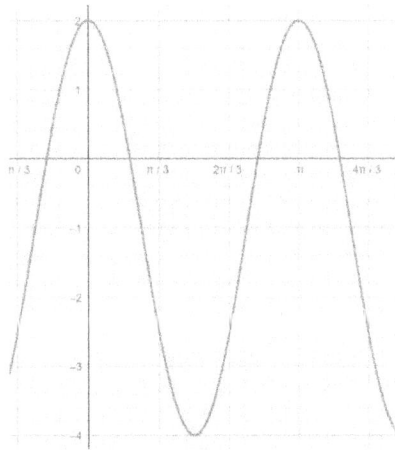

First, find the period and the value of B.

$$\pi = \frac{2\pi}{B}$$

So:

$$B = \frac{2\pi}{\pi} = 2$$

Next, find the midline. Here it is $y = -1$. So, we know the function has been shifted down 1.

Then, knowing the midline, find the amplitude. From the new midline, the curve goes up 3 and down 3. So, our amplitude is 3.

Finally, determine if this is a cosine function or a sine function. Cosine starts up (typically at 1) and sine (typically) starts at the origin. Here, this function starts up, so it is cosine.

Putting together the pieces, we have:

$$f(x) = 3\cos(2x) - 1$$

Work this problem. (Remember, there is not a phase shift.) Find an equation for the following graph:

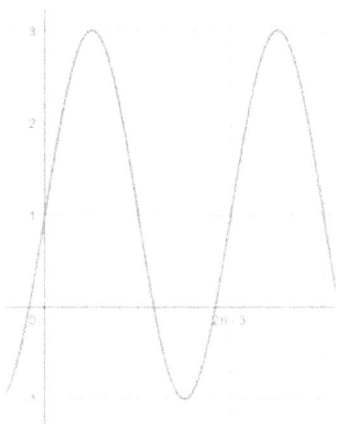

x) Period:

y) $B =$

z) Midline:

aa) Amplitude:

bb) Sine or Cosine:

cc) Give the full equation of the function:

Trig

Active Learning: 8.2a

Below is a graph of the tangent function: $f(x) = \tan x$

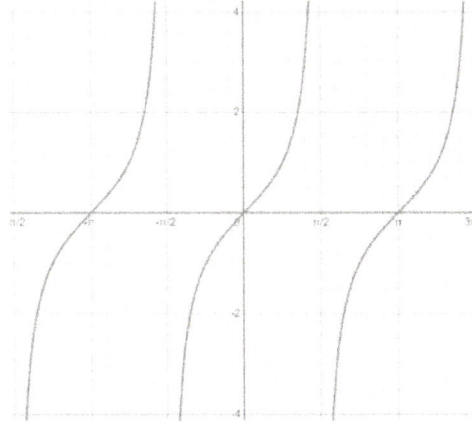

a) What is the period of the tangent function? (Remember, the period is how far on the x-axis it takes for the shape to repeat.)

b) The tangent function has vertical asymptotes. Think of $\tan x = \frac{\sin x}{\cos x}$. Look at the graph. Why does it have vertical asymptotes at the locations which it does?

c) The amplitude was the distance from the midline to the highest and lowest points. The sine and cosine functions have an amplitude, however the tangent function does not. Looking at the graph, why can't the tangent function have an amplitude?

The transformations of a tangent function follow the same essential ideas which we saw for sine and cosine, with only small changes. The amplitude is now just a vertical stretch (called the stretch factor) and the period is based on π. I'll walk through one:

$$f(x) = 3\tan(3x - \pi) + 1$$

First, pull out the value in front of the x as a GCF. This is to make sure we get the horizontal shift (phase shift) correct.

$$f(x) = 3\tan\left(3\left(x - \frac{\pi}{3}\right)\right) + 1$$

Stretch Factor: Vertical stretch of 3.

Period: We will start with π for a tangent function and divide it by the value in front of the x.

$$Period = \frac{\pi}{3}$$

Phase Shift: Right $\frac{\pi}{3}$

Midline: The midline was the x-axis ($y = 0$) and it has been shifted vertically up 1. So, the midline is $y = 1$.

Try these: (Don't forget to pull out the GCF.)

$$f(x) = 2\tan(3x - 2\pi) - 3$$

d) Stretch Factor:

e) Period:

f) Phase Shift:

g) Midline:

$$f(x) = 5\tan(2x - 3\pi) + 4$$

h) Stretch Factor:

i) Period:

j) Phase Shift:

k) Midline:

Next, we want to graph tangent functions. However, before we do, we need to look at three key points for a tangent function.

Tangent is a function. We put in the angle and get out the ratio. This creates coordinates we can use to graph the tangent function. Evaluate the following and give the coordinates. These points are selected because they are easy to work with.

l) $f(0) = \tan(0) =$

(0, __)

m) $f\left(\frac{\pi}{4}\right) = \tan\left(\frac{\pi}{4}\right) =$

$\left(\frac{\pi}{4}, \_\_\right)$

n) $f\left(-\frac{\pi}{4}\right) = \tan\left(-\frac{\pi}{4}\right) =$

$\left(-\frac{\pi}{4}, \_\_\right)$

The easiest way to graph a tangent function is to modify these three points based on the transformations. Let's look through an example.

$$f(x) = 3\tan(4x) + 1$$

We will start with the three key points and transform them one step at a time. Transformations work from the inside out and so we will do the same. First, we have a horizontal compression of 4. The x coordinate is the horizontal coordinate and a compression means division. So, we will divide the x coordinate by 4.

Dividing x by 4

$(0, 0) \to (0, 0)$

$\left(\frac{\pi}{4}, 1\right) \to \left(\frac{\pi}{16}, 1\right)$

$\left(-\frac{\pi}{4}, -1\right) \to \left(-\frac{\pi}{16}, -1\right)$

The asymptotes are also impacted by the horizontal compression, so we also divide them by 4.

$x = \frac{\pi}{2} \to x = \frac{\pi}{8}$

$x = -\frac{\pi}{2} \to x = -\frac{\pi}{8}$

Next, we have a vertical stretch of 3. A vertical stretch would impact the y coordinate. So, we multiply the y coordinate of our new points by 3.

        Multiplying the y coordinate by 3.

$(0, 0) \rightarrow (0, 0)$ (Multiplying 3 times zero is still zero.)

$\left(\frac{\pi}{16}, 1\right) \rightarrow \left(\frac{\pi}{16}, 3\right)$

$\left(-\frac{\pi}{16}, -1\right) \rightarrow \left(-\frac{\pi}{16}, -3\right)$

Finally, there is a vertical shift up 1. This would impact the y coordinate and so we add one to each y.

$(0, 0) \rightarrow (0, 1)$

$\left(\frac{\pi}{16}, 3\right) \rightarrow \left(\frac{\pi}{16}, 4\right)$

$\left(-\frac{\pi}{16}, -3\right) \rightarrow \left(-\frac{\pi}{16}, -2\right)$

Graphing gives us:

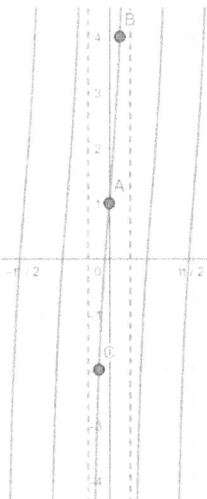

Try one.

Here is our function.

$$f(x) = 2\tan(3x) - 3$$

Your transformations are:

- A horizontal compression of 3. (So, you will divide the x-coordinates and the asymptotes by 3.)
- A vertical stretch of 2. (So, you will multiply the y-coordinate by 2.)
- A vertical shift down of 3. (So, you will subtract 3 from the y-coordinates.)

o) $(0, 0) \rightarrow$

p) $\left(\frac{\pi}{4}, 1\right) \rightarrow$

q) $\left(-\frac{\pi}{4}, -1\right) \rightarrow$

r) $x = \frac{\pi}{2} \rightarrow$

s) $x = -\frac{\pi}{2} \rightarrow$

t) You would then plot the points and the asymptotes to get the graph. Circle the correct graph below:

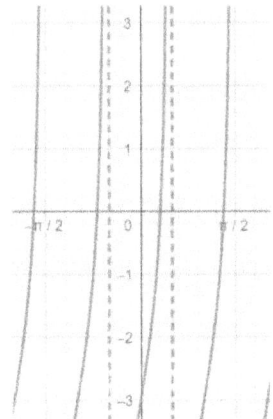 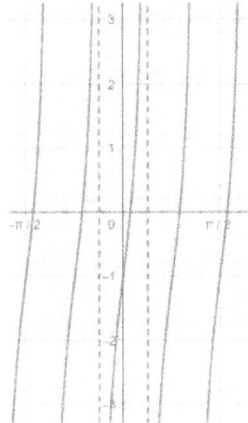

Adding horizontal shifts makes this harder and I do not ask for it in my classroom. However, it could be done by adding an additional step. Horizontal shifts would require moving the x-coordinate either left or right.

Trig

Active Learning: 8.2b

Just as the secant function is based upon cosine, the graph of the secant function can be made based upon the graph of the cosine function.

a) The graph of secant has vertical asymptotes wherever the graph of cosine hits the x-axis. On the graph below, draw vertical asymptotes (dashed lines). Why does secant have vertical asymptotes at these locations? (Hint: $\sec x = \frac{1}{\cos x}$)

Because secant is the reciprocal of cosine it goes up when cosine goes down and it goes down when cosine goes up. Graphing it is easy. At the top of every hill (maxima) draw a U which is pinned between the asymptotes. At the bottom of every valley (minima) draw a ∩ which is pinned between the asymptotes. Do so on the graph below.

b)

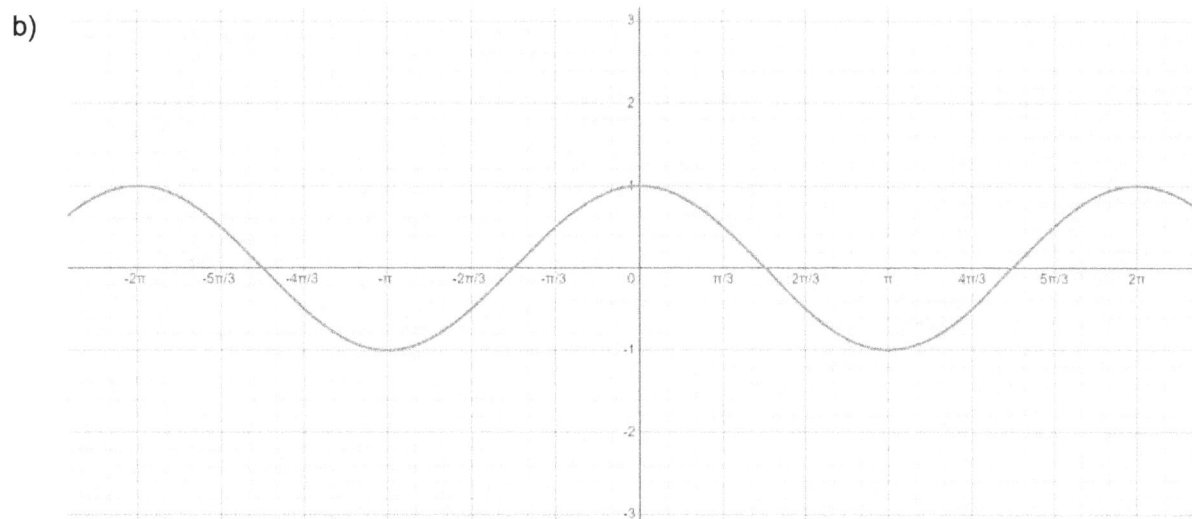

Here is a transformed secant function:

$$f(x) = 2\sec(2(x - \frac{\pi}{2}))$$

Describe the transformations:

c) Stretch: (No Amplitude for Secant)

d) Period: (Because it is based on cosine, we start with a period of 2π.)

e) Phase Shift:

f) Midline:

To graph the function, we start with

$$f(x) = 2\cos(2(x - \frac{\pi}{2}))$$

Add the asymptotes wherever the graph below hits the x-axis. Then add the U's at the tops of the peeks and valleys.

g)

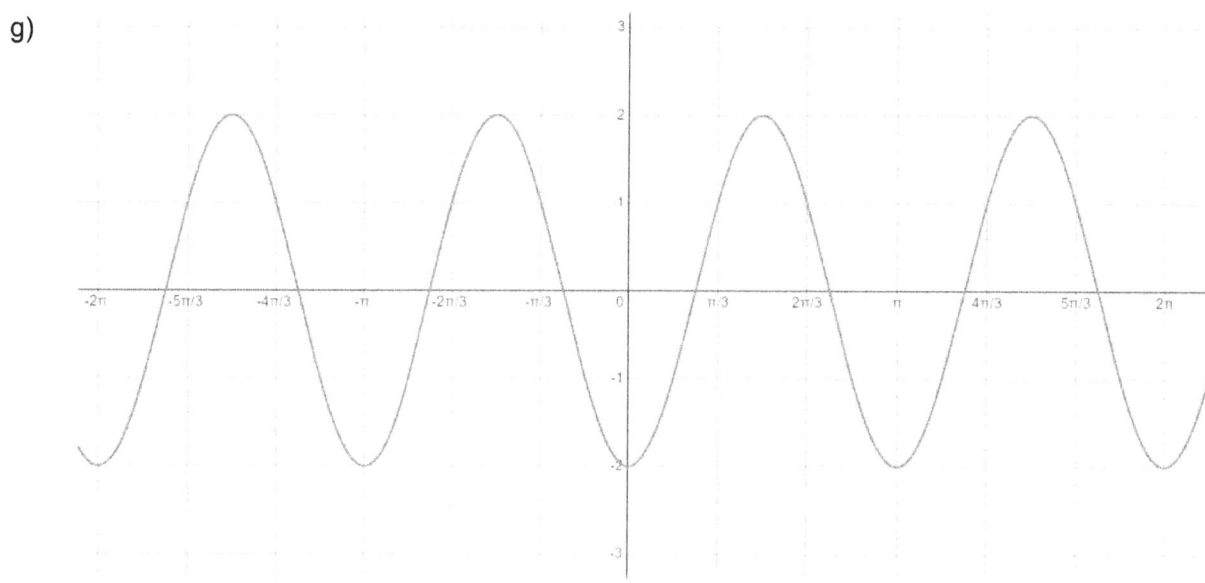

The concept is identical for cosecant. Below is the graph of sine. Add vertical asymptotes wherever the sine function hits the x-axis. Add U's at the peeks and valleys, bounded by the asymptotes.

h)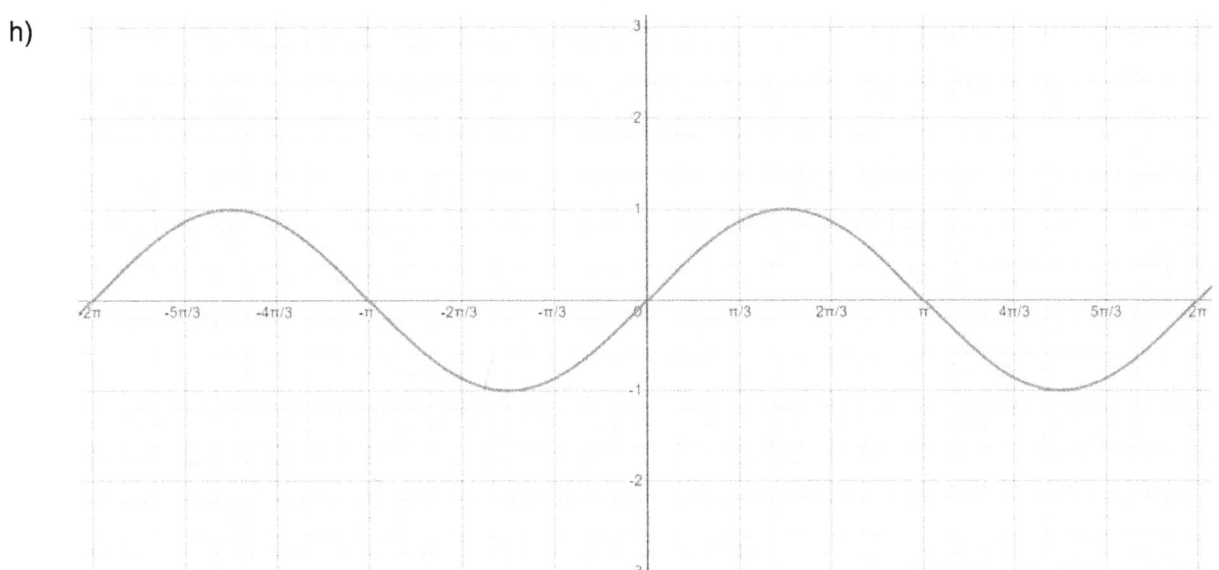

Here is a transformed cosecant function:

$$f(x) = 2\csc\left(3\left(x + \frac{\pi}{4}\right)\right)$$

Describe the transformations:

i) Stretch: (No Amplitude for cosecant)

j) Period: (Because it is based on sine, we start with a period of 2π .)

k) Phase Shift:

l) Midline:

To graph the function, we start with:

$$f(x) = 2\sin\left(3\left(x + \frac{\pi}{4}\right)\right)$$

Add the asymptotes wherever the graph below hits the x-axis. Then add the U's at the tops of the peeks and valleys.

m)

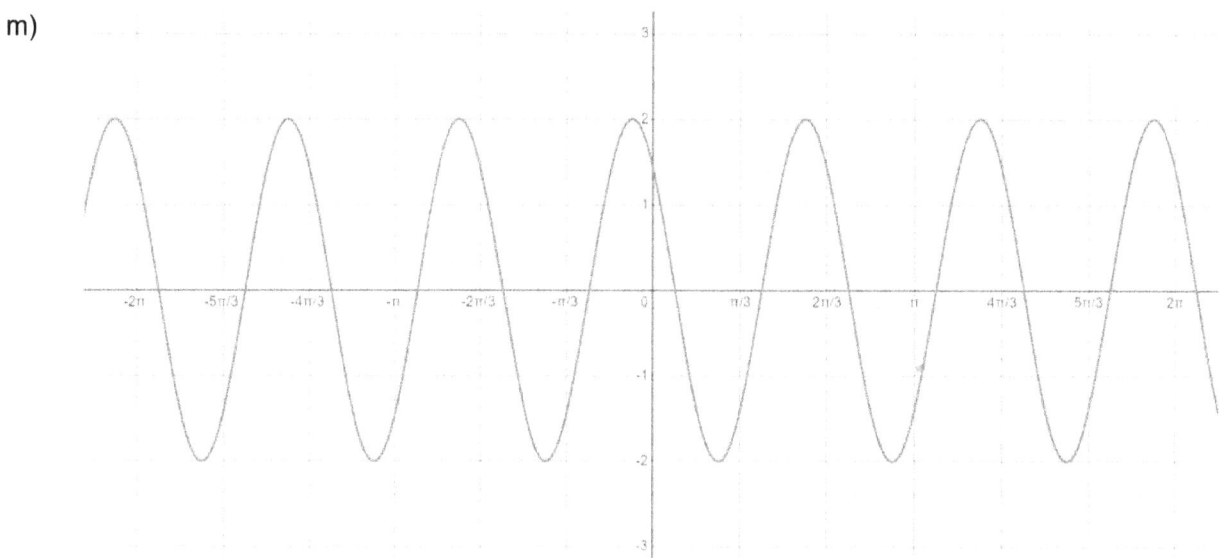

Trig

Active Learning: 8.3a

Today we need to recall inverse functions. They are function machines that work in reverse. To check to see if a function can have an inverse, we use a horizontal line. If a horizontal line touches the graph twice, it cannot have an inverse. If a horizontal line touches the graph only once, it can have an inverse.

a) Circle any graphs below which can have an inverse.

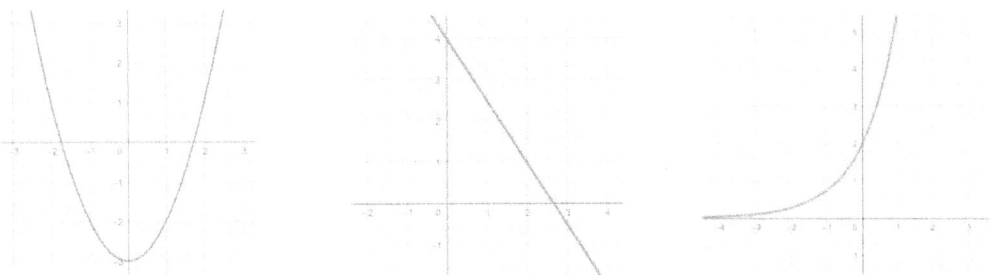

One of the functions below is the cosine function and one is the sine function. Label them and indicate if they are one-to-one.

b)

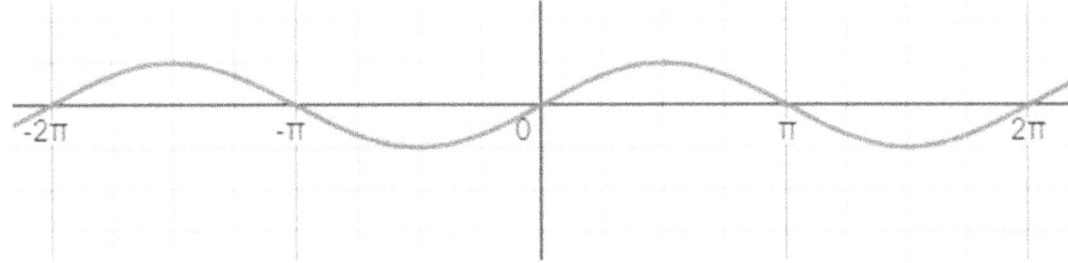

c)

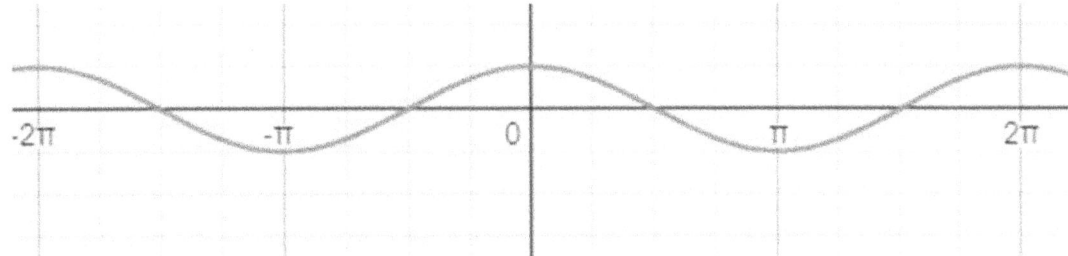

When a function is not one-to-one (and therefore can't have an inverse) there is a mathematical trick we can use. We can limit the domain. Basically, we cut the function so that it is one-to-one.

Here is how the sine function gets cut:

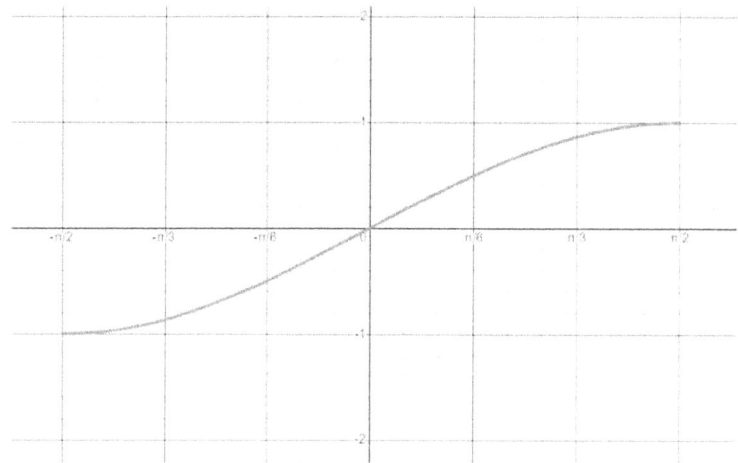

Give the domain and range of this new "cut" sine function. (Remember, domain is the possible x-values and range is the possible y-values.)

d) Domain:

e) Range:

Here is how the cosine function gets cut:

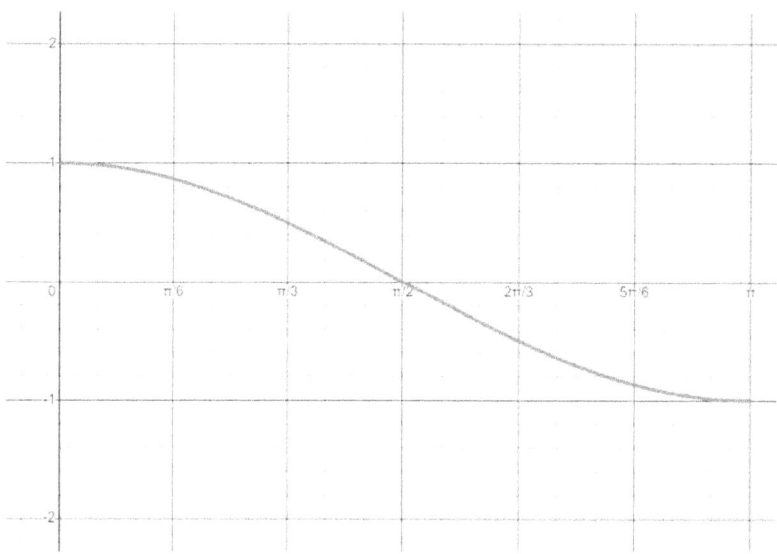

Give the domain and range of this new "cut" cosine function.

f) Domain:

g) Range:

The "cuts" of the sine and cosine function could be made in other locations, but these are the customary locations.

After the functions have been cut in this manner, they are now one-to-one and so they can have inverses. Here are their inverse functions.

$y = \sin^{-1}(x)$ (Sometimes written as $y = \arcsin(x)$)

$y = \cos^{-1}(x)$ (Sometimes written as $y = \arccos(x)$)

Give the domain and range of these functions. (Remember, they are inverse functions, so they "run" in reverse, flipping the domain and range.)

$$y = \sin^{-1}(x)$$

h) Domain:

i) Range:

$$y = \cos^{-1}(x)$$

j) Domain:

k) Range:

Next, let us evaluate some inverse functions.

$$y = \sin^{-1}\left(\frac{1}{2}\right)$$

l) This is an inverse function, so it works in reverse. So, the question being asked here is "What angles have a sine ratio of $\frac{1}{2}$?" (There are two. List them both.)

m) However, only one of these angles is within the range of the inverse sine function. Which angle is within the range of $\sin^{-1}(x)$?

$$y = \cos^{-1}\left(-\frac{1}{2}\right)$$

n) Again, this is an inverse function, so it works in reverse. So, what angles have a cosine ratio of $-\frac{1}{2}$?

o) However, only one of these angles is within the range of the inverse cosine function. Which angle is within the range of $\cos^{-1}(x)$?

We could also use a calculator to find inverse functions. But be careful. As we've seen before, your calculator can be set to either degrees or radians. Use radian mode unless the problem is clearly indicating degrees.

Find the value of the following:

p) $\cos^{-1}(-.6) =$

q) $\sin^{-1}(.19) =$

r) $\tan^{-1}(4) =$

In these next problems, use an inverse trig function to find the angle. You will need to choose the right inverse function based upon the sides which you have been given.

s)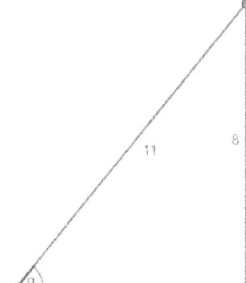

t)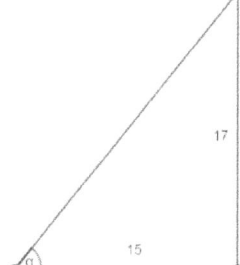

Trig

Active Learning: 8.3b

In Algebra II, we learn about composite functions: $f(g(x))$. Composite functions are like assembly lines. An input x goes into the function $g(x)$. The answer that comes out is then put directly into the function $f(x)$. We are going to create composites involving inverse trig functions.

There are two types. First, a composite where the inverse is on the outside (the second machine): $f^{-1}(g(x))$). Second, a composite where the inverse is on the inside (the first machine): $f(g^{-1}(x))$). The inverse on the outside is easier and we will start with that.

Composite trig functions of the type: $f^{-1}(g(x))$

1) $\cos^{-1}(\sin\frac{7\pi}{6})$

a) When the inverse is on the outside, just follow the assembly line. First, from your unit circle, what is the value of $\sin\frac{7\pi}{6}$?

b) Now we need to find: $\cos^{-1}(answer)$. There are two θ's which have this ratio of sine. What are they?

c) But $\cos^{-1}$ is limited to quadrants I and II. Which θ is in the appropriate quadrant?

2) $\sin^{-1}(\cos\frac{5\pi}{6})$

d) When the inverse is on the outside, just follow the assembly line. First, from your unit circle, what is the value of $\cos\frac{5\pi}{6}$?

e) Now we need to find: $\sin^{-1}(answer)$. There are two θ's which have this ratio of sine. What are they?

f) But $\sin^{-1}$ is limited to quadrants I and IV. Which θ is in the appropriate quadrant?

g) This problem has an extra twist. In that quadrant, sin^{-1} using negative angles. Which of the following is the correct angle:

   a) $-\frac{2\pi}{3}$
   b) $-\frac{\pi}{3}$
   c) $-\frac{\pi}{6}$

Composite trig functions of the type: $f(g^{-1}(x))$

$$\cos\left(\sin^{-1}\left(\frac{3}{5}\right)\right)$$

With composite functions we always start with the inside. But with the inverse on the inside, the idea is a little different. Build a triangle with a sine ratio of $\frac{3}{5}$.

h) Finish the triangle. Find the missing side by using the Pythagorean Theorem.

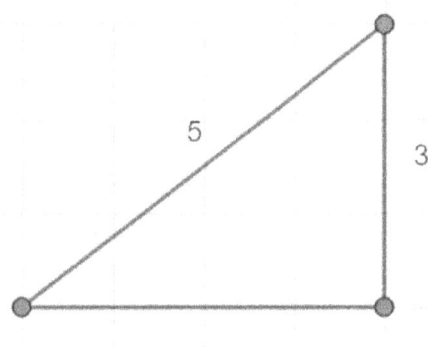

i) This triangle is the inside of the composite function. Now, use this triangle to find the value of cosine. What is the value of cosine for this triangle?

The inside inverse function builds the triangle and the outside trig function references that triangle to provide the final answer.

Try another.

$$\sin\left(\tan^{-1}\left(\frac{5}{12}\right)\right)$$

Again, the inverse trig function is on the inside. So, build a triangle with a tan ratio of $\frac{5}{12}$. I've set it up for you.

j)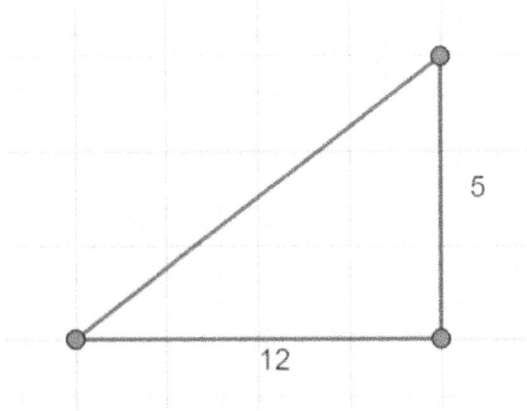

Finish the triangle by using the Pythagorean Theorem. What is the value of the hypotenuse?

k) This triangle is the inside of the composite function. Now, just use this triangle to find the value of sine. What is the value of sine for this triangle?

Trig

Active Learning: 8.3c

Here, we are going to look at a unique problem which is of the type: $f(g^{-1}(x))$. The problem is algebra-based and we don't have actual numbers.

$$\sin\left(\cos^{-1}\left(\frac{x}{x+1}\right)\right)$$

Since the inverse is on the inside, we build a triangle:

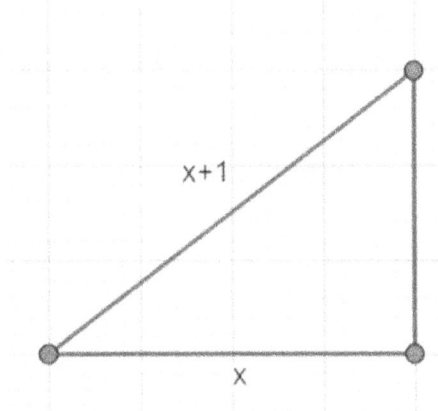

And, as before, we will use the Pythagorean Theorem to complete the triangle. Unlike previous problems, now we must use algebra.

$$x^2 + b^2 = (x+1)^2$$

a) Solve for the missing side, b. Be careful, $(x+1)^2 = (x+1)(x+1)$ and must be foiled.

b) With the triangle now complete, what is the value of sine?

Try one on your own.

$$\cos\left(\sin^{-1}\left(\frac{x}{1}\right)\right)$$

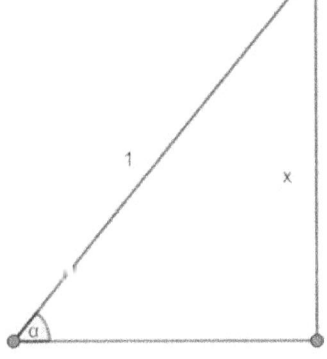

c) What is the value of the missing side?

d) Now, find the value of cosine.

Here is a variation of the same type of problem.

Find $\cos t$ if $\sin t = \dfrac{x}{\sqrt{x^2+1}}$.

Both trig functions are referring to the same angle. So, build a triangle off $\sin t$ and then use that triangle to find $\cos t$. I've gotten you started below. (You will again need to find the missing side.)

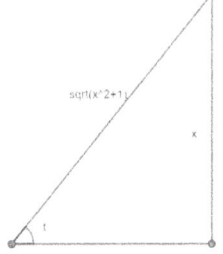

e) What is the value of the missing side?

f) What is the value of $\cos t$?

Trig

Active Learning: 9.1a

We are going to solve this system of linear equations by the substitution method.

$$2x + y = 0$$
$$3x + y = 1$$

a) First, solve the top equation for y. Write your answer in the space below.

b) Next, since you know an equivalent expression, you can replace y in the second equation. In the space below, replace y in the second equation and solve for x.

c) Finally, plug the value for x back into the first equation to find the value of y.

We want to begin a topic called trig identities. Trig identities look like equations, but they aren't. Rather, they are mathematical proofs. Our goal is to show that one side does, in fact, equal the other side. And, we will see that we have two primary tools to help us: algebra and substitution.

Let's solve a simple problem.

$$\cos(x)\tan(x) = \sin(x)$$

We know that $\tan(x) = \frac{\sin(x)}{\cos(x)}$.

d) Replace $\tan(x)$ with the equivalent expression. Then simplify to show that the left side is the same as the right side.

$$\cos(x)\tan(x) = \sin(x)$$

e) Now, let's do a slightly harder one. Substitute into the left side and use algebra to simplify.

$$\frac{\cot(x)}{\csc(x)} = \cos(x)$$

Use these facts.

$$\cot(x) = \frac{\cos(x)}{\sin(x)}$$

$$\csc(x) = \frac{1}{\sin(x)}$$

f) Next, we will use algebra before we substitute. Get a common denominator.

$$1 + \frac{\sin^2(x)}{\cos^2(x)} = \sec^2(x)$$

g) Then use the Pythagorean identity $\sin^2(x) + \cos^2(x) = 1$. Make your substitution and then finish proving the identity. Show your steps.

h) Try one with a bit less help. (Hint: In the numerator, you need to start by using a modified version of the Pythagorean identity.)

$$\frac{1 - \sin^2(x)}{\cos(x)} = \cos(x)$$

Again, it is important to remember that these are a type of mathematical proof. They aren't equations. For instance, I can't multiply $\cos(x)$ across the equal sign. We don't yet know this is truly equal. That is what we are trying to prove.

Finally, the textbook includes problem which are similar to but aren't actually identities. The following is an example.

Simplify the trig expression by writing the simplified form in terms of the second expression.

$$\frac{\tan x}{\cot x + \tan x} ; \sin x$$

This is asking us to write the expression on the left with only values of sine. Here are some tips:

- Substitute all the identities in terms of cosine and sine.
- When needed get a common denominator.
- Use the Pythagorean Identity.
- Keep, Change, Flip the bottom fraction.

i) Try the problem below:

Trig

Active Learning: 9.1b

a) One group of identities involves even and odd functions. As we learned earlier, cosine is an even function. Which of the following is true:

    a) $\cos(-x) = \cos(x)$
    b) $\cos(-x) = -\cos(x)$

b) And, sine is an odd function. Which of the following is true:

    a) $\sin(-x) = \sin(x)$
    b) $\sin(-x) = -\sin(x)$

Finish the following. (Remember, the key is whether or not they are based on sine or cosine.)

c) $\csc(-x) =$

d) $\sec(-x) =$

e) $\tan(-x) =$

f) $\cot(-x) =$

When working with identities, if you have a negative inside with the x, address it. Use the even/odd properties to remove the x from the inside.

Prove the following identity:

g) $\sin(-x)\cot(-x)\cos(-x) = \cos^2(x)$

h) For some identities, we first need to remember some factoring. Factor the following:

$$x^2 - 1$$

87

i) If a trig expression follows the same pattern, it can be factored too. Use the same idea to factor this:

$$tan^2(x) - 1$$

j) Now, prove the following identity. (Hint: Factor the numerator and pull a GCF out of the denominator.)

$$\frac{tan^2(x) - 1}{tan(x)\sin(x) + \sin(x)} = \sec(x) - \csc(x)$$

k) In the next identity, we need a trick that is common, but not very intuitive. Multiply the following: (Hint: you need to F.O.I.L.)

$$(1 + \sin x)(1 - \sin x) =$$

l) With the help of the Pythagorean Identity, turn your answer into something based on cosine.

m) Let's use that trick to prove this next identity. Work with the left and multiply the top and bottom of the fraction by $(1 - \sin x)$.

$$\frac{\cos x}{1 + \sin x} = \frac{1 - \sin x}{\cos x}$$

Trig

Active Learning: 9.2a

a) The unit circle is based (primarily) on three reference angles. What are those reference angles? (Give them in both degrees and radians.)

But what if we are interested in evaluating a trig function at an angle other than those reference angles? One trick is to recreate this unknown angle from the sum or difference of two angles from our unit circle.

Find two angles (from the unit circle) which could be added to make the following angle: (Call the first angle u and the second angle v. There may be more than one possibility.)

b) • 105°

$u =$

$v =$

c) • $\frac{5\pi}{12}$ (Hint: These are from the first quadrant.)

$u =$

$v =$

d) • 255°

$u =$

$v =$

Find two angles (from the unit circle) which could be subtracted to make the following angle: (Once again, call the first angle u and the second angle v.)

e)
- 105°

$u =$

$v =$

f)
- $\frac{\pi}{12}$

$u =$

$v =$

g)
- 165°

$u =$

$v =$

If we can turn an angle whose value we don't know into a sum or difference of two angles that we do, we can use the following formulas to evaluate it.

$$\sin(u + v) = \sin u \cos v + \cos u \sin v$$
$$\sin(u - v) = \sin u \cos v - \cos u \sin v$$

Find the sine of the following angles. I've given you a head start.

h) $\sin(165°) = \sin(120° + 45°) =$

i) $\sin\left(\dfrac{7\pi}{12}\right) = \sin\left(\dfrac{5\pi}{6} - \dfrac{\pi}{4}\right) =$

$$\cos(u + v) = \cos u \cos v - \sin u \sin v$$
$$\cos(u - v) = \cos u \cos v - \sin u \sin v$$

Find the cosine of the following angles.

j) $\cos(165°) = \cos(120° + 45°) =$

k) $\cos\left(\dfrac{7\pi}{12}\right) = \cos\left(\dfrac{5\pi}{6} - \dfrac{\pi}{4}\right) =$

l) Suppose we have two angles: u and v. The $\sin u = -\frac{4}{5}$ and is in Quadrant III. The $\cos v = -\frac{5}{8}$ and is in Quadrant II. On the grid below, draw triangles that represent each of the angles. (Remember, you know two of the sides.)

Find the following:

m) $\cos u =$

n) $\sin v =$

Now that you know the sine and cosine of both angles, you could evaluate their sum or difference. Find $\cos(u - v)$. (Notice that you needed the sine and cosine for both angles for the formulas.)

$$\cos(u - v) = \cos u \cos v - \sin u \sin v$$

o) $\cos(u - v) =$

Trig

Active Learning: 9.2b

As we saw in the last lesson, we use sum or difference formula to help us find the exact value of a trig ratio that we couldn't otherwise find. Just as sine and cosine have sum or difference formulas, so does tangent.

$$\tan(\alpha + \beta) = \frac{\tan\alpha + \tan\beta}{1 - \tan\alpha \tan\beta} \qquad \tan(\alpha - \beta) = \frac{\tan\alpha - \tan\beta}{1 + \tan\alpha \tan\beta}$$

a) Which two first quadrant angles sum to make $\frac{5\pi}{12}$:

    a) $\frac{\pi}{6} + \frac{\pi}{4}$

    b) $\frac{\pi}{3} + \frac{\pi}{4}$

    c) $\frac{\pi}{6} + \frac{\pi}{3}$

So, $\tan\left(\frac{5\pi}{12}\right) = \tan\left(\frac{\pi}{6} + \frac{\pi}{4}\right)$.

For the sum formula, we will need the following values of tangent: $\tan\frac{\pi}{6}$ and $\tan\frac{\pi}{4}$. To find them, use the unit circle to put sine over cosine. Simplify the fractions but don't rationalize the denominator. In the sum and difference formulas it is easier to save that until all the math has been done.

b) $\tan\frac{\pi}{6} = \dfrac{\sin\frac{\pi}{6}}{\cos\frac{\pi}{6}} =$

c) $\tan\frac{\pi}{4} = \dfrac{\sin\frac{\pi}{4}}{\cos\frac{\pi}{4}} =$

Now, substitute those values into the formula to find the sum. Simply by getting one fraction on top and one on bottom. You will get a fraction over a fraction. Keep Change Flip.

d) $\tan\left(\dfrac{5\pi}{12}\right) = \tan\left(\dfrac{\pi}{6} + \dfrac{\pi}{4}\right) = \dfrac{\tan\frac{\pi}{6} + \tan\frac{\pi}{4}}{1 - \left(\tan\frac{\pi}{6}\right)\left(\tan\frac{\pi}{4}\right)}$

Earlier, we learned about cofunctions. Cofunctions are partner functions which have the same value when they add to 90° (or $\frac{\pi}{2}$). Match the trig function on the left with his cofunction on the right:

e) $\cos x$ $\hspace{4cm}$ $\tan x$

f) $\sec x$ $\hspace{4cm}$ $\sin x$

g) $\cot x$ $\hspace{4cm}$ $\csc x$

Finish each of the following. Give a cofunction which would have the same value.

h) $\cos 72° =$

i) $\sec \dfrac{\pi}{7} =$

j) $\cot \dfrac{\pi}{9} =$

Whenever we have something unusual in a trig identity, we should begin by substituting it out. Here is an identity which starts by having a difference of a sine.

$$\frac{\sin(\alpha - \beta)}{\cos \alpha \cos \beta} = \tan \alpha - \tan \beta$$

k) Prove this identity. Follow these two tips:

1) Replace $\sin(\alpha - \beta)$ with the difference formula.
2) Split what you now have into two separate fractions, both over $\cos \alpha \cos \beta$.

Trig

Active Learning: 9.3a

In this activity, we are going to look at more formulas which address special situations involving angles.

Double-Angle Formula

To better understand, we need to start with building a triangle. $\sin \alpha = \frac{5}{8}$, and it is in the second quadrant.

a)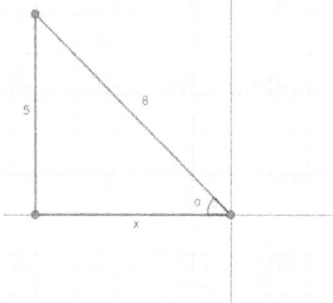

Finish the triangle using the Pythagorean Theorem. (We are in the second quadrant, so x will be negative.)

Use the triangle to find $\cos \alpha$.

b) $\cos \alpha =$

This triangle gives us information about the angle α. But what if we wanted information about an angle 2α that we know nothing about. (We know an angle and want information about an angle that is double.) The following identities can help us.

$$\sin(2\alpha) = 2\sin(\alpha)\cos(\alpha)$$
$$\cos(2\alpha) = \cos^2(\alpha) - \sin^2(\alpha)$$
$$\cos(2\alpha) = 1 - 2\sin^2(\alpha)$$
$$\cos(2\alpha) = 2\cos^2(\alpha) - 1$$

They are called double angle formulas. We can use the top formula to find $\sin(2\alpha)$ since our triangle gives us $\sin \alpha$ and $\cos \alpha$. List the values from the triangle again.

c) $\sin \alpha =$

d) $\cos \alpha =$

Now, plug those value into the double angle formula to find the value of $\sin(2\alpha)$.

e) $\sin(2\alpha) =$

There are three versions of the cos(2α) formula. They will all get you to the same place. Use any of the three to find cos(2α).

f) $\cos(2\alpha) =$

Try another. If $\tan \alpha = -\frac{3}{4}$ and it is in the second quadrant. Find:

(Don't forget to consider their signs.)

g)  $\sin(\alpha) =$

h) $\cos(\alpha) =$

Again, now that you know sine and cosine, you have everything you need to find an angle which is double in size. Use the formulas to find the following:

i) $\sin(2\alpha) =$

j) $\cos(2\alpha) =$

There is a double angle formula for tangent as well.

$$\tan(2\alpha) = \frac{2\tan(\alpha)}{1 - \tan^2(\alpha)}$$

Use the same triangle to find tan(2α).

k) $\tan(2\alpha) =$

Double angle problems can come in lots of variations. For instance, you could be asked to work them backward.

Simplify to one trig expression.

$$8 \sin\left(\frac{\pi}{8}\right) \cos\left(\frac{\pi}{8}\right)$$

If we pull out a 4, we have something which follows the double angle formula.

$$4\left(2 \sin\left(\frac{\pi}{8}\right) \cos\left(\frac{\pi}{8}\right)\right)$$

Using the double angle formula backward allows us to simplify to one trig expression.

$$4\left(2 \sin\left(\frac{\pi}{8}\right) \cos\left(\frac{\pi}{8}\right)\right) = 4 \sin\left(2 \cdot \frac{\pi}{8}\right)$$

l) I showed the double angle formula being taken backward. Now, in the space below, simplify the angle inside the parenthesis.

Try one on your own.

m) Simplify to one trig expression.

$$6\cos^2\left(\frac{\pi}{8}\right) - 3$$

n) Another problem variation would involve using a double angle to solve an identity. Prove the following identity:

$$\frac{\sec(\alpha) \sin(2\alpha)}{2} = \sin(\alpha)$$

Begin by making this substitution: $\sin(2\alpha) = 2 \sin(\alpha) \cos(\alpha)$.

Trig

Active Learning 9.3b

It can be useful to get rid of (reduce) the exponents on a trig function. The following formulas can be used on even powers of sine, cosine, or tangent. As you can see, they involve double angles.

$$\sin^2(\alpha) = \frac{1 - \cos(2\alpha)}{2}$$

$$\cos^2(\alpha) = \frac{1 + \cos(2\alpha)}{2}$$

$$\tan^2(\alpha) = \frac{1 - \cos(2\alpha)}{1 + \cos(2\alpha)}$$

Here's an example of a problem. Rewrite the expression with an exponent no higher than one.

$$\cos^2(4x)$$

a) Use the power-reducing formula. (Hint: Your new angle will be $8x$.)

$\cos^2(4x) =$

Problems get more involved when they ask you to reduce a higher power. I don't include problems like this in my trig classes, but I will show the idea.

Rewrite the expression with an exponent no higher than one.

$$\sin^4(6x)$$

So, an equivalent version of this expression would be:

$$(\sin^2(6x))^2$$

And, using the power reducing formula, we get:

$$\left(\frac{1 - \cos(12x)}{2}\right)^2 = \left(\frac{1 - \cos(12x)}{2}\right)\left(\frac{1 - \cos(12x)}{2}\right)$$

b) When multiplying fractions, we multiply the top times the top and the bottom times the bottom. Try that here. (Hint: The numerators follow the same pattern as if you were multiplying $(1-w)(1-w)$.)

101

c) Your answer should now include one term with an exponent of two. That needs to be reduced again. Try below. (I've shown the final answer.)

The answer should be: $\frac{3}{8} - \frac{1}{2}\cos(12x) + \frac{1}{8}\cos(24x)$

d) Finally, work an identity involving the power-reducing formulas. It looks difficult, but is actually rather straightforward.

$$\cos^3(4x) = \frac{1}{2}\cos(4x)(1 + \cos(8x))$$

I'll get you started with the first step.

$$\cos^3(4x) = (\cos(4x))(\cos^2(4x))$$

Trig

Active Learning: 9.3c

Half-angle Formulas

We will approach half-angle formulas in a similar way to which we did the double-angle formulas. First, let's build a triangle. $\sin \alpha = -\frac{4}{5}$, and it is in the fourth quadrant.

a) Finish the triangle using the Pythagorean Theorem.

Use the triangle to find $\cos \alpha$.

b) $\cos \alpha =$

This triangle gives us information about the angle α. But what if we wanted information about an angle $\frac{\alpha}{2}$ that we know nothing about. (We know an angle and want information about an angle that is half.) The following identities can help us.

$$\sin\left(\frac{\alpha}{2}\right) = \pm\sqrt{\frac{1-\cos(\alpha)}{2}}$$

$$\cos\left(\frac{\alpha}{2}\right) = \pm\sqrt{\frac{1+\cos(\alpha)}{2}}$$

When the angle we are interested in is half of one we know, we can use the half-angle formulas. Notice that both formulas only require $\cos \alpha$. Find the sine of the half angle.

c) $\sin\left(\frac{\alpha}{2}\right) =$

d) But the formula has a twist. We took a square root and square roots can be $\pm$. So, we need to know the proper sign. The sign is determined by the quadrant of the half-angle. If the full-sized angle α was in the fourth quadrant, and the half-angle is (obviously) half, what quadrant must the half-angle be in?

Find the value of sine in that quadrant and you have the sign you need for your answer. With that in mind, what is the proper value of:

e) $\sin\left(\dfrac{\alpha}{2}\right) =$

Using the same triangle, find the value of $\cos\left(\dfrac{\alpha}{2}\right)$. (Be sure to give the proper sign to the final answer.)

f) $\cos\left(\dfrac{\alpha}{2}\right) =$

Here's a slightly different version of a problem which requires the half-angle formula. Find $\cos 15°$.

We don't know anything about 15°. But we know something about 30°. Here is the triangle for 30° inside of the unit circle. (We can't build a triangle as we did in the first problem, but we don't need to since we know the values of sine and cosine.)

g)

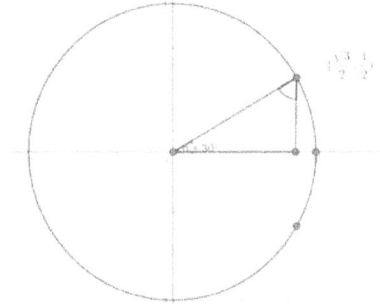

What is the value of $\cos 30°$?

So, we know something about 30° and we want to find $\cos 15°$. 15° is half of 30°. Therefore, we can use the half-angle formula.

Use the half-angle formula to find $\cos 15°$. (Since we know 15° is in the first quadrant, your answer should be positive.)

h) $\cos 15° =$

Let's work a similar problem in radians.

Find the exact value using half-angle formulas.

$\cos\left(\frac{\pi}{8}\right) =$

This is an angle we know nothing about, but it is half of an angle we know. Look at our formula:

$$\cos\left(\frac{\alpha}{2}\right) = \pm\sqrt{\frac{1+\cos(\alpha)}{2}}$$

Let $\frac{\alpha}{2} = \frac{\pi}{8}$. So, $\alpha = \frac{2\pi}{8} = \frac{\pi}{4}$.

So, if we know α, and this is an angle from the unit circle, we can use our half-angle formula.

$$\cos\left(\frac{\pi}{8}\right) = \pm\sqrt{\frac{1+\cos\left(\frac{\pi}{4}\right)}{2}}$$

Finish the formula to find the value of $\cos\left(\frac{\pi}{8}\right)$. (The sign will be positive because $\frac{\pi}{8}$ is in Quadrant I.)

i) $\cos\left(\frac{\pi}{8}\right) =$

Try one on your own.

Find the exact value using half-angle formulas.

j) $\sin\left(\frac{5\pi}{12}\right) =$

(Your answer will be positive here because $\frac{5\pi}{12}$ is in the first quadrant. But remember to always consider the proper sign.)

There are also versions of the half-angle formula for tangent. There are three versions, however here is one which is derived simply by putting the sine and cosine half-angle formulas over top of one another.

$$\tan\left(\frac{\alpha}{2}\right) = \pm\sqrt{\frac{1-\cos\alpha}{1+\cos\alpha}}$$

Use the half-angle formula for tangent to solve the following problem.

Find the exact value using half-angle formulas.

k) $\tan\left(-\dfrac{\pi}{8}\right) =$

(Notice that this is in quadrant IV. Tangent is negative there and so your answer should choose the negative sign.)

The use of the double-angle and half-angle formulas gets confusing. Here's my suggestion on how to remember. I know something about an angle, but I don't know about the half-angle—use the half-angle formula. I know something about an angle, but I don't know about the double-angle—use the double angle formula.

Trig

Active Learning: 9.4a

a) In this activity, we continue with formulas which can modify a trig expression. Here, we want to look at how to change the product of two trig functions into a sum or difference. The proofs for these formulas are pretty straightforward. They are derived by adding or subtracting two of the sum or difference identities. Work the first proof below by adding the sum and difference formulas for cosine. (Hint: These formulas have been proven, so you can move across the equal sign. You are trying to create the first formula from the list below.)

$$\cos\alpha\cos\beta + \sin\alpha\sin\beta = \cos(\alpha - \beta)$$
$$+ \cos\alpha\cos\beta - \sin\alpha\sin\beta = \cos(\alpha + \beta)$$

Here is a list of all the product to sum formulas:

$$\cos\alpha\cos\beta = \frac{1}{2}[\cos(\alpha - \beta) + \cos(\alpha + \beta)]$$

$$\sin\alpha\sin\beta = \frac{1}{2}[\cos(\alpha - \beta) - \cos(\alpha + \beta)]$$

$$\sin\alpha\cos\beta = \frac{1}{2}[\sin(\alpha + \beta) + \sin(\alpha - \beta)]$$

$$\cos\alpha\sin\beta = \frac{1}{2}[\sin(\alpha + \beta) - \sin(\alpha - \beta)]$$

Write the following products as sums.

b) $\sin\left(\dfrac{8x}{3}\right)\sin\left(\dfrac{5x}{3}\right) =$

c) $4\sin(6\theta)\cos(2\theta) =$

On this next problem, you will create a negative angle. Use the even/odd identities to simplify it further.

d) $8 \cos(2\theta) \sin(7\theta) =$

After using the product to sum formula, this problem can be evaluated using values from the unit circle.

e) $\dfrac{1}{2} \cos\left(\dfrac{13\pi}{12}\right) \cos\left(\dfrac{\pi}{12}\right) =$

Trig

Active Learning: 9.4b

In the last activity, we learned how to use identities to change products to sums and differences. Here, we will change sums and differences into products. The proof isn't difficult. However, it requires a clever substitution and so we will not work it. Below are the formulas:

$$\sin \alpha + \sin \beta = 2 \sin\left(\frac{\alpha + \beta}{2}\right) \cos\left(\frac{\alpha - \beta}{2}\right)$$

$$\cos \alpha - \cos \beta = -2 \sin\left(\frac{\alpha + \beta}{2}\right) \sin\left(\frac{\alpha - \beta}{2}\right)$$

$$\sin \alpha - \sin \beta = 2 \sin\left(\frac{\alpha - \beta}{2}\right) \cos\left(\frac{\alpha + \beta}{2}\right)$$

$$\cos \alpha + \cos \beta = 2 \cos\left(\frac{\alpha + \beta}{2}\right) \cos\left(\frac{\alpha - \beta}{2}\right)$$

Write the following sums or differences as a product.

a) $\sin 8\theta - \sin 2\theta =$

(This problem will create a negative angle. Use the even/odd identities to simplify to a positive angle.)

b) $\cos 3\theta + \cos 7\theta =$

In this next problem, the product formula will include a value which you can evaluate using the unit circle.

c) $\sin(15°) + \sin(75°) =$

Trig

Active Learning: 9.5a

Provide the values of the following:

a) $\sin(30°) =$

b) $\sin(390°) =$

c) $\sin(750°) =$

d) $\sin(-330°) =$

e) What is happening? Explain why.

Provide the values of the following:

f) $\tan(45°) =$

g) $\tan(225°) =$

h) $\tan(405°) =$

i) $\tan(585°) =$

With the tangent function things are slightly different, and it has to do with the period of the function.

j) What is happening? Explain why.

Below is a trigonometric equation, which I have solved.

$2\cos(x) = 1$

$\cos(x) = \dfrac{1}{2}$

$x = \dfrac{\pi}{3} + (2\pi)k$ where k equals any integer.

Or

$x = \dfrac{5\pi}{3} + (2\pi)k$ where k equals any integer.

k) The $\dfrac{\pi}{3}$ and $\dfrac{5\pi}{3}$ are the two angles where cosine equals $\dfrac{1}{2}$, but why has $+(2\pi)k$ been added? (Hint: Think circle.)

l) Solve the following trigonometric equation. Give your answer in radians. (Be sure to include $+(2\pi)k$.)

$2\sin(x) = \sqrt{3}$

Here, I have worked a trigonometric function involving tangent.

$4\tan(x) = -2\sqrt{2}$

$\tan(x) = \dfrac{-2\sqrt{2}}{4}$

$\tan(x) = -\dfrac{\sqrt{2}}{2}$

m) $x = \dfrac{3\pi}{4} + n\pi$ where n equals any integer. Why does my answer now include $+n\pi$ instead of $+(2\pi)n$?

112

n) Solve the following equation.

$$x = \sqrt{2} - x$$

When solving equations with trig functions, they work just like variables. Solve the following and isolate the trig functions just as you would if you were solving for a variable. But once you have sine alone, you will also need to do as we did above and find the angles which have that sine ratio. (Don't forget your $+(2\pi)k$.)

o)
$$\sin(x) = \sqrt{2} - \sin(x)$$

If you are solving a trig equation and the problem says $0 \leq x < 2\pi$, then the answer only involves the original circle and you don't need to add the $+(2\pi)k$.

Work the following. You don't need $+(2\pi)k$.

p) Solve: $2\cos\theta - 3 = -5, 0 \leq \theta < 2\pi$.

q) Solve: $2\sin x + 1 = 0, 0 \leq x < 2\pi$.

If you need to solve a trig equation by using a calculator, be careful. Remember, that an inverse function has been "cut" in order for there to be an inverse function. When you work inverse trig functions on your calculator it will only give values in the following quadrants:

- $sin^{-1}x$ will return values in Quadrants I and IV.

- $cos^{-1}x$ will return values in Quadrants I and II.

- $tan^{-1}x$ will return values in Quadrants I and IV.

r) Use the inverse function on a calculator to find $sin(\theta) = .1$. Assume that θ is in radians.

The answer you have found is in Quadrant I. However, sine is also positive in quadrant II, but your calculator doesn't provide that answer. To find the value in quadrant II, think of your answer as a reference angle.

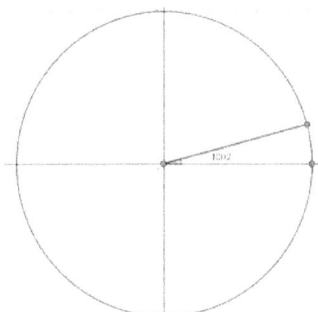

s) Here is the angle being used as a reference angle in quadrant II. We need to subtract from the nearest x-axis (which in radians has an angle of π.) Find the value of $sin(\theta) = .1$ in quadrant II.

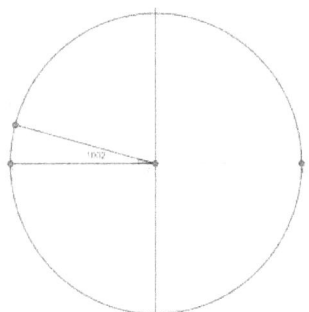

114

Try another. This time with cosine. (Remember, your calculator will give inverse values of cosine in quadrants I and II.)

t) $\cos^{-1}(.2) =$ (Assume that θ is in radians.)

u) Cosine is also positive in Quadrant IV. Use reference angles to find the value in quadrant IV.

Trig

Active Learning: 9.5b

The next idea is not hard to do, but it is hard to understand. Look at this problem:

$$\cos(2\theta) = \frac{\sqrt{3}}{2}$$

Let's solve it just as we would if it wasn't 2θ. There are two angles that have a sign value of $\frac{\sqrt{3}}{2}$.

$$\cos(2\theta) = \frac{\sqrt{3}}{2}$$

$$2\theta = \frac{\pi}{6} \text{ and } 2\theta = \frac{11\pi}{6}$$

Now, we will solve each of those values for θ.

$$2\theta = \frac{\pi}{6}$$

Find θ.

a) $\theta =$

Next, as we typically do, add 2π to your answer for θ.

b) $\theta =$

Notice that the value you just created is coterminal but it is not a unique value between $0 \leq \theta < 2\pi$.

Solve the second value for θ.

$$2\theta = \frac{11\pi}{6}$$

Find θ.

c) $\theta =$

Now add 2π to your answer.

d) $\theta =$

Once again, this value you just created is coterminal. However, it is not a unique value between $0 \leq \theta < 2\pi$.

But try this value for θ.

$$\theta = \frac{13\pi}{12}$$

$$\cos\left(2 \cdot \frac{13\pi}{12}\right) = \frac{\sqrt{3}}{2}$$

e) Is this true?

And try this value for θ.

$$\theta = \frac{23\pi}{12}$$

$$\cos\left(2 \cdot \frac{23\pi}{12}\right) = \frac{\sqrt{3}}{2}$$

f) Is this true?

Here's what we are seeing. $\theta = \frac{13\pi}{12}$ and $\theta = \frac{23\pi}{12}$ are unique values between $0 \leq \theta < 2\pi$. And they are also values which make the equation true. So, what is going on? Why can't we simply add 2π to our answers like we did before.

g) To help us, solve the following for y:

$$2y = 4x + 8$$

Here, we know that we must divide the entire right side by 2. In the same way suppose the following is true.

$$2\theta = \frac{\pi}{6} + 2\pi k$$

If we solve for θ, we must divide the entire right side by 2.

$$2\theta = \frac{\pi}{6} + 2\pi k$$

And we would get:

$$\theta = \frac{\pi}{12} + \pi k$$

118

h) Add the following:

$$\frac{\pi}{12} + \pi =$$

i) Solve our second solution for θ, and remember to divide the entire right side by 2:

$$2\theta = \frac{11\pi}{6} + 2\pi k$$

$\theta =$

j) Add π to your solution for θ.

$\theta + \pi =$

This is where I found the other unique solutions to the equation. So, when we divided the "spin around the circle" by 2, we found two additional answers. Dividing angles always creates this. It is hard to wrap your head around, but mathematically we can see it is true. Therefore, if we must divide angles, we must also divide the $+2\pi k$. If we don't, we will miss some solutions.

Try the following: (Don't forget to divide the $+2\pi k$.)

k) $\cos(2\theta) = \frac{1}{2}$

l) $\sin(3\theta) = \frac{\sqrt{3}}{2}$

Trig

Active Learning: 9.5c

a) For our next idea, we need a refresher on an old one. Solve the following:

(Hint: First, pull out a GCF. Then, use the zero-product property to solve for the two solutions.)

$2x^2 + x = 0$

As we've seen before, trig functions work exactly the same as variables. Follow the exact same pattern as you did above to factor and solve the trig equation below:

(Hint: Unlike in the problem above, once you know what the two sine functions equal, you must find the angle θ which make that equation true. There is also no restriction which indicates only values of θ for one trip around the circle. So, be sure to include $+2\pi k$.)

b) $2\sin^2\theta + \sin\theta = 0$

This next problem doesn't require factoring. However, it has solutions which are not on the unit circle and will require a calculator. I'll show the first few steps for you.

$$6\sin^2\theta = 2$$

$$\sin^2\theta = \frac{1}{3}$$

$$\sin\theta = \pm\sqrt{\frac{1}{3}}$$

$$\sin\theta = \pm\frac{\sqrt{3}}{3}$$

$$\sin\theta = \pm.5774$$

c) Use the inverse sine function on your calculator to find the values of θ which make this true. You will need to work it twice, once for the positive value and once for the negative. What are your two values of θ?

d) But, if you recall, inverse sine only gives values in the first and fourth quadrant. You also have values in the second and third quadrant. Use reference angles to give those additional two answers.

e) For the next problem, let's return to factoring. Factor the problem below and then solve.

(Hint: The a/c method is easiest.)

$$2x^2 - 5x + 3 = 0$$

f) Follow the same pattern as you did with x to solve for $\sin \theta$. (Don't take the last step of solving for θ yet.)

$$2\sin^2 \theta - 5 \sin \theta + 3 = 0, 0 \leq \theta < 2\pi$$

g) Recall that the sine function has a range of $[-1, 1]$. So, one of your equations has a value of sine which is outside of sine's range. Since this equation is impossible, we simply disregard it. Solve the other equation for θ. Notice that you are restricted to one time around the circle.)

h) Try another. Once again, start with an equivalent version of the problem involving x.

$$12x^2 - 15x = -9x$$

i) Now solve the trig version. The pattern is still the same.

(Hint: The two equations you get for sine are both valid, so solve them both for θ. Notice the restriction on one time around the circle.)

$$12\sin^2 x - 15 \sin x = -9 \sin x, 0 \leq x < 2\pi$$

j) Work one final problem involving tangent. There is no factoring. However, you will need to use your calculator. The necessary values of tangent aren't on the unit circle.

(Hint: Since you take a square root a $\pm$ is involved. You also should get four unique answers.)

$$2 \tan^2 \theta - 3 = 0, 0 \leq x < 2\pi$$

Trig

Active Learning: 10.1a

In this activity, we want to investigate an idea called The Law of Sines. It is a tool for working with obtuse (non-right) triangles. Use the Triangle below to answer the questions.

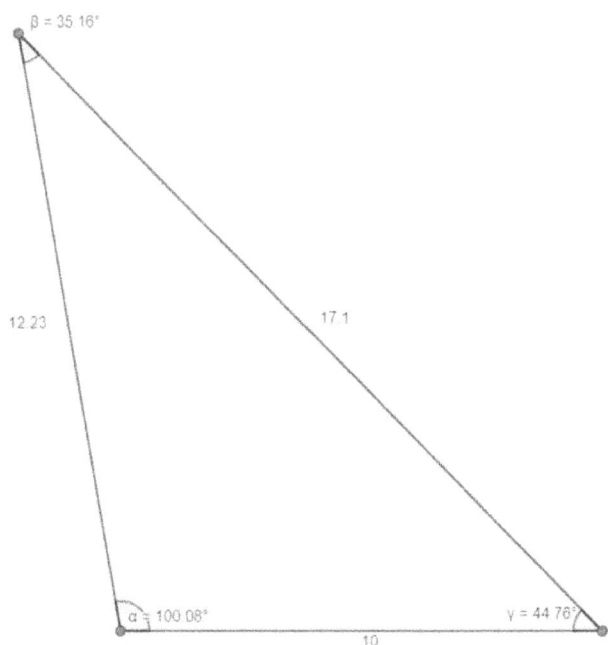

Find the following ratios:

a) $\dfrac{17.1}{\sin(100.08)} =$

b) $\dfrac{12.23}{\sin(44.76)} =$

c) $\dfrac{10}{\sin(35.16)} =$

d) What is true about these ratios?

e) What is the relationship between the side and angle that were chosen for each ratio?

Find the value of the ratios if they are calculated as reciprocals.

f) $\dfrac{\sin(100.08)}{17.1} =$

g) $\dfrac{\sin(44.76)}{12.23} =$

h) $\dfrac{\sin(35.16)}{10} =$

i) What is true about these ratios?

There are two versions of The Law of Sines. Below, I have removed a mathematical operator from the equations. Fill in The Law of Sines with the missing symbols.

j) (Hint: What is true about each of the ratios? Give the mathematical symbol which states that.)

$$\dfrac{\sin \alpha}{a} \qquad \dfrac{\sin \beta}{b} \qquad \dfrac{\sin \gamma}{c}$$

$$\dfrac{a}{\sin \alpha} \qquad \dfrac{b}{\sin \beta} \qquad \dfrac{c}{\sin \gamma}$$

To use The Law of Sines, we need one complete ratio (one angle and its corresponding side) and either another side or angle. (The equation simply says that all three ratios are the same, but we only use two of the ratios to work a problem.) Look at the triangle below:

k)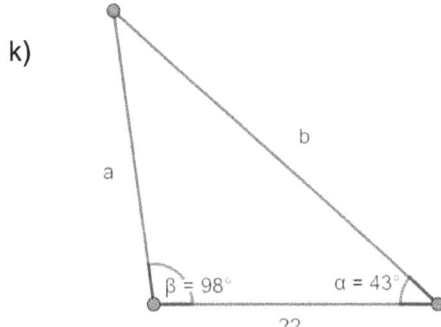

We need the angle γ to have one complete ratio. But since we know the other two angles inside a triangle, we can find it. What is the measure of angle γ?

Use The Law of Sines to find the values of the missing sides. (Notice that you can't use the Pythagorean theorem because it is not a right triangle.)

l) $a =$

m) $b =$

Try one more. Find the missing sides of the triangle.

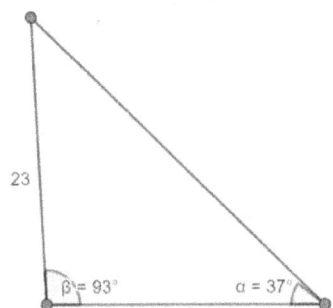

n) $b =$

o) $c =$

Trig

Active Learning: 10.1b

As I mentioned previously, to use The Law of Sines, we must have one complete ratio. A SSA triangle is one where we have two consecutive sides and then a single angle. The angle we have *is* across from one of the sides and so we have a complete ratio. But since we don't know all of the angles of the triangle, we can't be sure what the triangle looks like. There are three possibilities, which we will look at in this activity.

Outcome #1

Use The Law of Sines to work the problem below. Something will go wrong. Be sure to note what went wrong and explain what it means.

a) A triangle has an angle of 50° with the side opposite being equal to 4. A second side has a length of 10. Find all the angles and sides of the triangle.

b) What went wrong? Why? (Hint: It has to do with the range of sine.)

Option #2

The following proportion has been set up using The Law of Sines. Use your calculator to find angle α.

c) $\dfrac{120}{\sin 80°} = \dfrac{121}{\sin \alpha}$

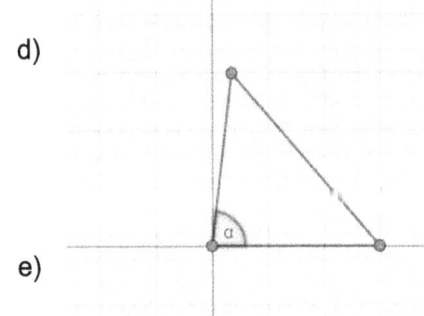

d) We already knew one angle in this triangle was 80°, now that we know a second, it is easy to find the third. What is the third angle in this triangle?

e) Using The Law of Sines, find the remaining missing side.

However, there is more to this problem than at first sight. Earlier we learned that when your calculator does inverse sine, it only gives answers in quadrants I and IV. Sine is also positive in quadrant II. Using our answer for angle α as a reference angle creates another triangle.

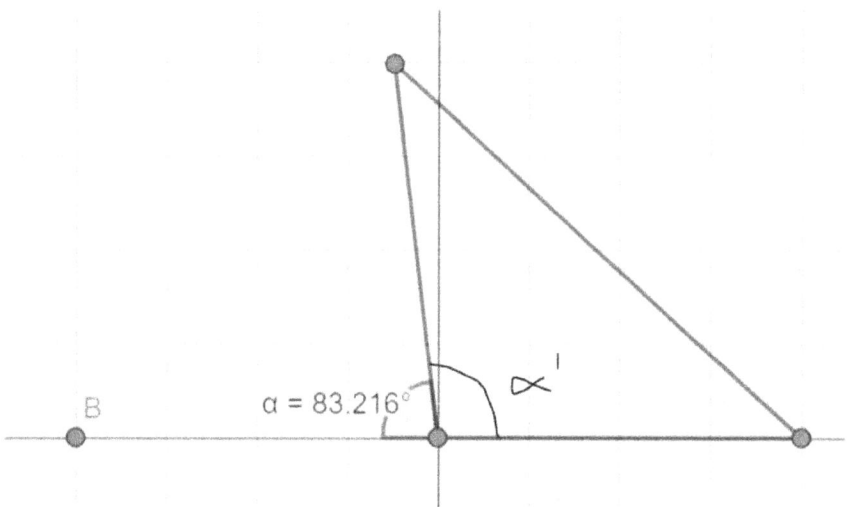

f) What is the value of α'?

So, if we have SSA when using The Law of Sines, we have a difficulty. Because we only know one angle and inverse sine could have two values, there may be a second triangle.

g) What would be the value of the third angle in this alternate triangle?

$\gamma' =$

This triangle also has an alternative length for the third side. Find it.

h) $c' =$

Option #3

There is one more possibility. Suppose we had the following:

$$\frac{35}{\sin 25°} = \frac{20.5}{\sin \alpha}$$

i) Solve for α.

We need to check for the possibility of a second triangle. We will find α' by subtracting our answer for α from 180°.

j) $\qquad\qquad\qquad\qquad 180° - \alpha =$

Now we know two of the angles in this possible second triangle. Find the value of the third angle in this triangle.

k) Something went wrong. Explain.

So, with SSA and The Law of Sines, we run into three possibilities: there are no possible triangles, two possible triangles, or just one. Next, let's look at the area of an obtuse triangle.

The Area of an Obtuse Triangle

Look at this triangle.

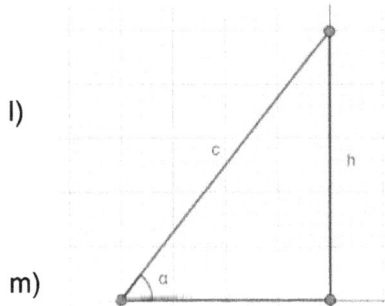

l) What is the value of the sine ratio for angle α? (There are no numbers, just use the variables.)

$\sin \alpha =$

Solve this equation for h.

m) $h =$

n) What is the formula for the area of a triangle?

Our goal is to find the area of an obtuse (non-right) triangle like the one below if we don't know the height.

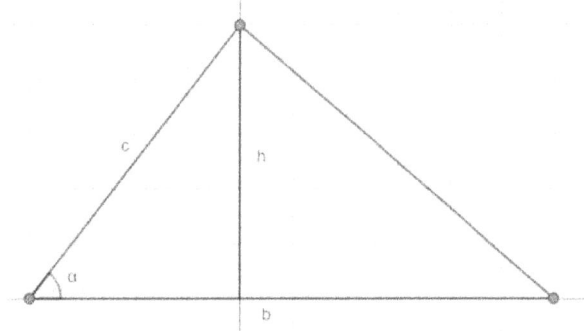

Notice the triangle in bold.

o)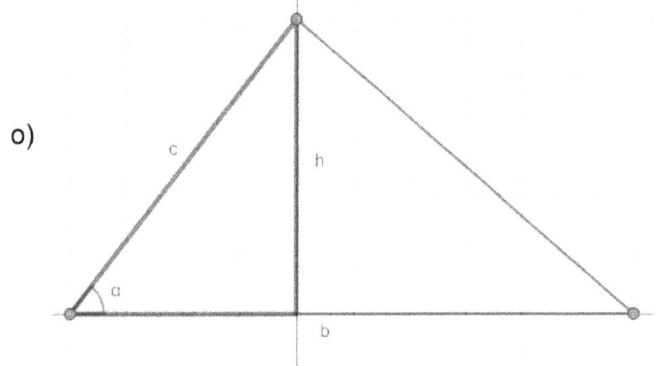

Earlier we used this same triangle to find a formula for h. What was that formula?

$h =$

In the area formula, remove h and replace it with our formula for h.

$$Area = \frac{1}{2} b \cdot h$$

p) $$Area = \frac{1}{2} b \cdot (\underline{\hspace{1cm}})$$

We now have a formula for the area of an obtuse triangle when the height is not known. All that is needed is an angle and the two adjacent sides.

q) Find the area of the following triangle:

$$a = 90$$
$$b = 52$$
$$\text{angle } C = 102°$$
$$Area = \frac{1}{2} a \cdot b \cdot \sin C$$

Trig

Active Learning: 10.2a

When we don't have a right triangle, the Law of Sines was helpful for finding the values of missing sides or angles. However, the Law of Sines required that we could set up one complete ratio. We needed at least one side and its corresponding angle. If we don't have that ratio, we need to use the Law of Cosines.

Below are three equivalent versions of the Law of Cosines.

Law of Cosines

1) $a^2 = b^2 + c^2 - 2bc \cos A$
2) $b^2 = a^2 + c^2 - 2ac \cos B$
3) $c^2 = a^2 + b^2 - 2ab \cos C$

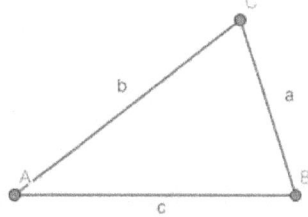

a) Look at the triangle on the right. In each version of the formula, there is a relationship between the side on the left of the equation and the angle needed on the right side of the equation. What is that relationship?

Next, we will use the Law of Cosines to solve a problem. Notice we are missing side a. Use the appropriate version of the Law of Cosines to find the value of a.

b) Solve the triangle with $A = 47°, b = 23$, and $c = 31$. (Round to 1 decimal place.)

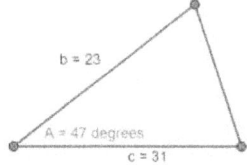

c) Now that you have side a, you have a complete ratio and can switch back to the Law of Sines. Use the Law of Sines to find angle B.

$$\frac{a}{\sin A} = \frac{b}{\sin B}$$

d) Finally, you know two of the three angles of a triangle. Use what you know about the sum of the angles in a triangle to find the measure of angle C.

Sometimes we have all three sides of a triangle, but we don't have any of the angles. In such circumstances, we can rearrange the Law of Cosines to solve for the angle.

Below, solve the three versions of the Law of Cosines for $\cos A$, $\cos B$, and $\cos C$.

e) (1)

f) (2)

g) (3)

Now, let's solve a problem where we have all three sides but are missing the angles.

Solve the triangle where $a = 22$, $b = 25.8$, and $c = 28.9$. (Round to the nearest tenth.)

h) Start by finding angle A. (Once you have the equation down to $\cos(A)$ you will use inverse cosine on your calculator.)

i) Next, find angle B. You have a ratio now, so if you prefer, you can switch to the Law of Sines. Or, you can once again use the Law of Cosines.

j) Finally, you have two of the three angles, use the sum of the angles of a triangle to find the third.

k) When we used our calculator to find inverse sine we were limited to which two quadrants?

l) Here, we are using inverse cosine. In which two quadrants are inverse cosine limited to?

Because inverse cosine returns a unique angle between 0° and 180°, we don't have to worry about the possibility of a second triangle.

137

Trig

Active Learning: 10.2b

In our last activities, we've learned how the Law of Sines and the Law of Cosines can help us to find the missing angles and sides of an obtuse (non-right) triangle. In this activity, we want to look at a loosely connected idea called Heron's Formula.

When you know the three sides of an obtuse triangle, we can use Heron's Formula to find the area.

$$Area = \sqrt{s(s-a)(s-b)(s-c)}$$

The s in the formula stands for the "half perimeter," where the three sides of the triangle have been added and then divided by 2.

$$s = \frac{(a+b+c)}{2}$$

The proof of the formula is clever math manipulation and isn't that intuitive, so I'll leave it to a textbook. However, the problems are pretty straightforward.

Use Heron's Formula to find the area of the obtuse triangle.

a)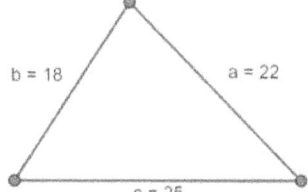

First, find the value of s by adding the three sides and dividing by 2.

$s =$

Now that you have the value of s, find the area. It is straightforward substitution and number crunching.

b) $Area =$

Try one more. Use Heron's Formula to find the area of a triangle with sides $a = 23.2, b = 29.5$, and $c = 34.1$.

c) $s =$

d) $Area =$

139

Trig

Active Learning: 10.3a

a) When plotting, we naturally think of a grid system, because we have been doing it this way our entire lives. On the graph below, plot the point (0, -5).

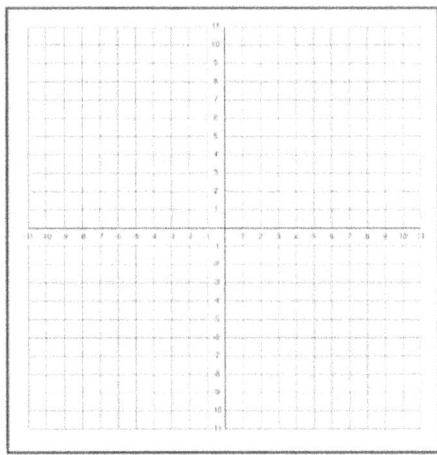

But, graphing doesn't have to be a grid system. Sometimes a circular system would make sense too. (Imagine dropping a rock into a pond.)

Our points are (r, θ), where r represents a particular circle from the center and θ represents the angle. The angles are measured just like they are in the unit circle.

b) On the graph, plot the point $(5, \frac{3\pi}{2})$.

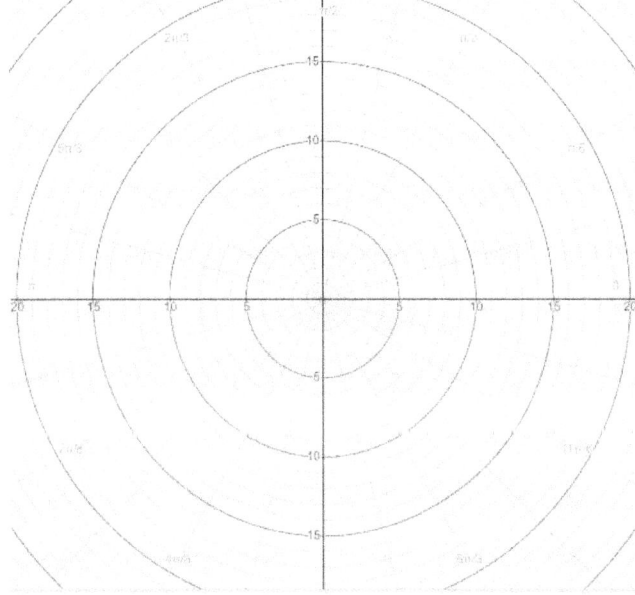

Next, let's learn how to switch between the two systems. Like most things we've seen, the key is a triangle.

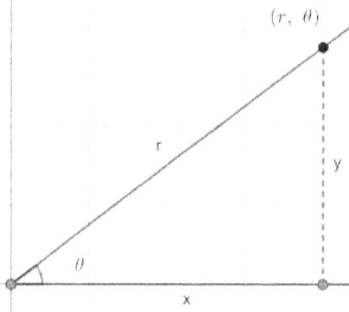

Notice, r is the hypotenuse of the triangle. So, find the values of our two primary trig functions in terms of x, y, and r.

c) $\cos \theta =$

d) $\sin \theta =$

Now, in the space below, solve each of those trig functions for either x or y.

e) $x =$

f) $y =$

g) If you had the values of x and y, but you didn't have r, what is the name of the formula which would allow you to find it?

h) Show that equation in terms of x, y, and r.

Now, we will use those formulas to convert from Polar form to Rectangular form. Find x and y, and you have the rectangular coordinates: (x, y). I will show the first problem as an example.

Convert $(5, \frac{3\pi}{2})$ to rectangular form. (Remember $r = 5$ and $\theta = \frac{3\pi}{2}$.)

$$x = r \cos \theta = 5 \cos \frac{3\pi}{2} = 5(0) = 0$$

$$y = r \sin \theta = 5 \sin \frac{3\pi}{2} = 5(-1) = -5$$

$$(0, -5)$$

i)   Convert $(3, \frac{5\pi}{6})$ to rectangular form.

j)   Convert $(4, \frac{7\pi}{4})$

Converting in the other direction, from Rectangular to Polar, is going to require the solving for r using the Pythagorean Theorem and tangent. Let me help you through one.

Convert $(-2, 2)$ to polar form.

First, let's find r.

$$r^2 = (-2)^2 + (2)^2$$

Find r. (Normally, if you take a square root, it is $\pm$. Here, however, r is the hypotenuse and is therefore always positive.)

k) $r =$

To find the angle we are going to use tangent.

$$\tan\left(\frac{y}{x}\right) = \tan\left(\frac{2}{-2}\right) = \tan(-1)$$

l)   What angle has a tangent of -1? (We know we are in the second quadrant from the original point.)

m)   Write the point in the form (r, θ)?

Try a couple on your own:

n) Convert the point $(-4, -4)$ to polar form.

o) Convert the point $(-2, -2\sqrt{3})$

Trig

Active Learning: 10.3b

In our last activity, we learned how to change points from Cartesian form (Rectangular) to Polar form (Circular). We are now going to do the same conversion to equations. The tools we need are these:

$$x^2 + y^2 = r^2$$
$$x = r \cos \theta$$
$$y = r \sin \theta$$

On the left side of these equations, we have the rectangular form and on the right side we have polar. We can use these to substitute between systems. Here is an example.

Convert the following equation to polar form.

$$x^2 + y^2 = 25$$

We can substitute r for $x^2 + y^2$.

$$x^2 + y^2 = 25$$
$$r^2 = 25$$

Finally, we will solve the equation for r.

$$r = \pm 5$$

So, $x^2 + y^2 = 25$ in rectangular form is equivalent to $r = \pm 5$ in polar form. They each create the graph of a circle with radius 5.

Try one on your own. I've given you one extra step, but otherwise it is the same.

Convert the following equation to polar form.

a) $x^2 = -y^2 + 121$

Here is a more complex problem.

$$x^2 + y^2 = 3x$$

There are two substitutions to make.

$$r^2 = 3r \cos \theta$$

The trick now is to solve for r.

$$r^2 - 3r\cos\theta = 0$$

b) The key is to factor out a Greatest Common Factor of r. Pull out the GCF of r and set each factor equal to zero.

One of your equations should be $r = 0$. This equation is simply a point at the origin and we can disregard it.

Try another. Convert the equation to polar form.

c) $x^2 + y^2 = 11y$

Work a couple more. On this first problem you will need to use both $x = r\cos\theta$ and $y = r\sin\theta$. (Hint: To solve for r, get all the terms with r on the left side and then pull out r as a GCF.)

d) $y = 5x + 3$

e) $y^2 = 4x - x^2$

Trig

Active Lesson: 10.3c

In the last activity, we converted rectangular equations to polar form. Now, we want to go the other direction, and it is a bit more difficult. We are going to use the same tools to make changes, solved slightly differently:

$$x^2 + y^2 = r^2$$

$$\frac{x}{r} = \cos\theta$$

$$\frac{y}{r} = \sin\theta$$

The first step in the conversion should be to eliminate any trig functions. Then, simplify before working with r.

Convert the following polar equation to rectangular form.

$$2r = 10\csc\theta$$

First, get rid of the trig function.

$$2r = 10 \cdot \frac{1}{\sin\theta}$$

Use the substitution for $\sin\theta$.

$$2r = 10 \cdot \frac{1}{\left(\frac{y}{r}\right)}$$

a) In this problem, rearranging will cause r to drop out. Finish by solving for y.

Try a similar problem on your own. Convert to rectangular form.

b) $3r = 15\sec\theta$

Let's look at some more difficult problems.

Convert to rectangular form.

$$r = \frac{5}{2 - 3\sin\theta}$$

As we did before, substitute out the trig function.

$$r = \frac{5}{2 - 3\left(\frac{y}{r}\right)}$$

We won't simplify the fraction. Instead, we will multiply up to the other side.

$$r\left(2 - 3\left(\frac{y}{r}\right)\right) = 5$$

c) To finish, distribute the r and then get r alone.

d) Finally, use the substitution $r = \sqrt{x^2 + y^2}$ and then square both sides.

To graph the equation, you would need to solve for y, but for our purpose you can leave the equation as it is. Try another similar problem.

Convert to rectangular form.

e) $r = \dfrac{6}{1 + 2\cos\theta}$

Let's look at one more.

Convert to rectangular form. First, substitute out the trig functions. Then, multiply across to the other side and the algebra will take care of the r^2.

f) $r^2 = 2 \csc \theta \sec \theta$

Trig

Active Learning: 10.5a

a) You have frequently graphed sets of ordered pairs on the coordinate plane using the x-axis and the y-axis. Plot the point $(-2, 3)$ on the graph below.

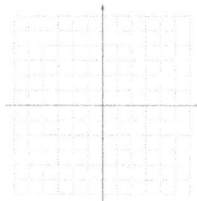

Recall, a complex number takes the form: $a + bi$. Where a is the real portion of the number and b is the imaginary portion. Mathematicians extended the idea of the coordinate plane to create something called a complex plane. The difference is that the x-axis is now the real portion of a complex number and the y-axis is now the imaginary portion of the complex number.
Here are two examples.

$-2 + 3i$ $3 - 4i$

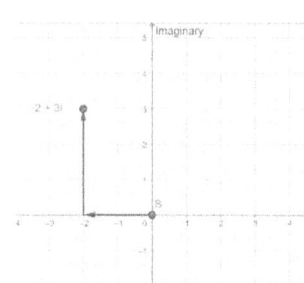 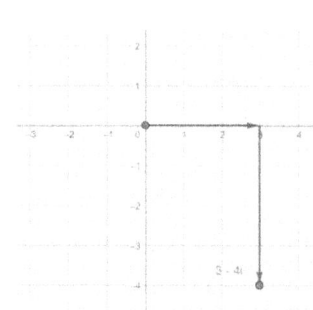

Plot the following numbers on the complex plane:

b) $5 + 2i$ c) $-4 - 2i$

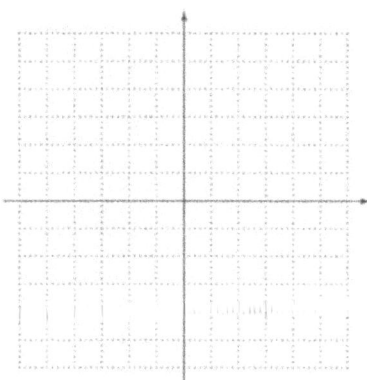 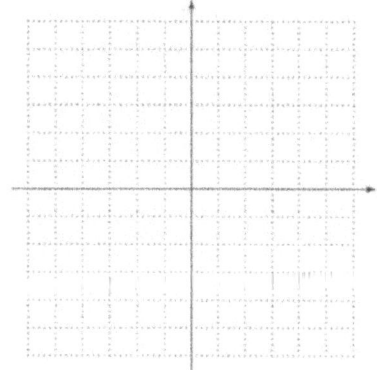

151

Converting a complex number into polar form is based on the same concept we saw in the last lab.

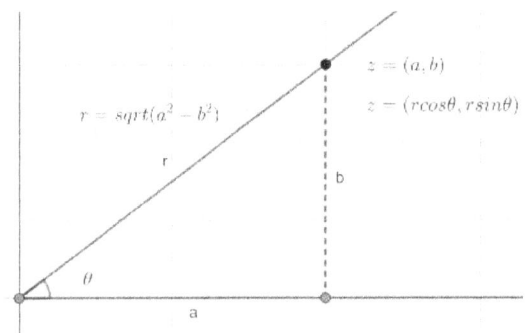

The value of r is found from the Pythagorean Theorem and is called the modulus. Here we always take the positive value. Find the modulus of the following complex numbers:

$3 + 2i$

d) $r =$

$5 - 4i$

e) $r =$

$-2 - 6i$

f) $r =$

We also need the value of θ. To find it, we use the tangent, like we did in the last section. (Remember that the imaginary portion of the number is the y and the real portion of the number is the x.)

Find the value of θ for the following complex numbers:

$2 - 2i$

$$\tan \theta = \tan\left(\frac{-2}{2}\right) = \tan(-1)$$

(Notice that you need an angle in quadrant IV because of the location of the original complex number.)

g) $\theta =$

$$1 + \sqrt{3}i$$

(Here, your angle should be in quadrant I.)

h) $\theta =$

The equation for the polar form is written below.

$$z = (r\cos\theta) + i(r\sin\theta) = r(\cos\theta + i\sin\theta)$$

So, to write a complex number in polar form, we find r and θ and then plug into the formula. (We are making the same substitutions as we did in the last activity.) Find the polar form of the following complex numbers:

$$-4 + 4i$$

i) $r =$

j) $\theta =$

k) $z =$

$$\sqrt{3} + i$$

l) $r =$

m) $\theta =$

n) $z =$

Converting from the Polar form of a complex number into the rectangular form is very straightforward. Simply find the value of the trig functions and then distribute r.

$$z = 4(\cos 120° + i\sin 120°)$$

$$4(-\frac{1}{2} + i\frac{\sqrt{3}}{2})$$

$$= -2 + 2\sqrt{3}i$$

Try a couple on your own. Convert from polar form into complex form.

o) $$z = 12(\cos\frac{\pi}{6} + i\sin\frac{\pi}{6})$$

p) $$z = 4(\cos\frac{5\pi}{3} + i\sin\frac{5\pi}{3})$$

Trig

Active Learning: 10.5b

In this lesson, we will begin by multiplying complex numbers in polar form. The equation turns out to be quite straightforward. Multiply the values of r and add the angles.

$$z_1 z_2 = r_1 r_2 [\cos(\theta_1 + \theta_2) + i \sin(\theta_1 + \theta_2)]$$

a) In the space below, prove this formula. You will need the following.

$$i^2 = -1$$

$$\cos(\theta_1)\cos(\theta_2) - \sin(\theta_1)\sin(\theta_2) = \cos(\theta_1 + \theta_2)$$

$$\sin(\theta_1)\cos(\theta_2) + \cos(\theta_1)\sin(\theta_2) = \sin(\theta_1 + \theta_2)$$

$$r_1(\cos(\theta_1) + i \sin(\theta_1)) \cdot r_2(\cos(\theta_2) + i \sin(\theta_2))$$

Multiply the following complex numbers. Again, simply multiply their values of r and add their angles. Problems may ask you to leave the multiplied values in polar form or to simplify down to rectangular form. For these problems, leave them in polar form.

$$z_1 = 4\left(\cos\left(\frac{4\pi}{3}\right) + i \sin\left(\frac{4\pi}{3}\right)\right) \text{ and } z_2 = 2\left(\cos\left(\frac{\pi}{2}\right) + i \sin\left(\frac{\pi}{2}\right)\right)$$

b) $z_1 z_2 =$

$$z_1 = 4(\cos(15°) + i \sin(15°)) \text{ and } z_2 = 12(\cos(285°) + i \sin(285°))$$

c) $z_1 z_2 =$

Division works in a similar fashion. (Although, I won't ask you to prove this one.) Divide the values of r and subtract the angles.

$$\frac{z_1}{z_2} = \frac{r_1}{r_2}[\cos(\theta_1 - \theta_2) + i\sin(\theta_1 - \theta_2)]$$

Try these. Leave your answers in polar form.

d) $\quad z_1 = 2(\cos\left(\frac{\pi}{2}\right) + i\sin\left(\frac{\pi}{2}\right))$ and $z_2 = 4(\cos\left(\frac{4\pi}{3}\right) + i\sin\left(\frac{4\pi}{3}\right))$

e) $\quad z_1 = 4(\cos(15°) + i\sin(15°))$ and $z_2 = 12(\cos(285°) + i\sin(285°))$

Exponents follow the same logic.

f) If $z = r(\cos\theta + i\sin\theta)$ show that $z^2 = r^2(\cos 2\theta + i\sin 2\theta)$.

(Hint: it is pretty easy. Use the formula for multiplying complex numbers and the fact that $\theta_1 = \theta_2$.)

It doesn't matter what power you need; the idea remains the same. It is called De Moivre's Theorem:

$$z^n = r^n[\cos(n\theta) + i\sin(n\theta)]$$

Find the value of the following and leave it in polar form:

g) $$[4(\cos(70°) + i\sin(70°))]^3$$

Our last problem has one additional step. The complex number is not in polar form, but to utilize De Moivre's Theorem it must be. First change it to polar form and then use the theorem to raise it to the fifth power.

h) $$(1-i)^5$$

Trig

Active Learning: 10.5c

In the last activity, we learned how to use De Moivre's Theorem to take a complex number in polar form and raise it to a power. In this activity, we are going to do the opposite and work with roots. When you studied Algebra I, you learned that the following are the same:

$$\sqrt{x} = x^{\frac{1}{2}}$$

a) If De Moivre's Theorem is for exponents, let's see what would happen if we put in a fractional exponent. In the formula below, replace n with $\frac{1}{n}$.

$$z^n = r^n[\cos(n\theta) + i\sin(n\theta)]$$

b) By entering $\frac{1}{n}$ into the formula, you have created $\frac{\theta}{n}$. We learned earlier that when we divide an angle additional unique solutions are created. To find them, we must also divide $2\pi k$. I've added $+2\pi k$ to De Moivre's Theorem. Once again, replace n with $\frac{1}{n}$, but this time, also divide $2\pi k$ by n.

$$z^n = r^n[\cos(n\theta + 2\pi k) + i\sin(n\theta + 2\pi k)]$$

The formula you have created placed fractional exponents into De Moivre's Theorem, but fractional exponents are roots. So, we now have a formula for taking roots of complex numbers in polar form.

c) Find the square root of $z = 36(\cos\frac{7\pi}{6} + i\sin\frac{7\pi}{6})$.

d) If we take a square root, there will be two complex roots. To find them, we will start with $k = 0$. Enter 0 for k into the formula to find the first solution.

e) The second solution will occur when $k = 1$. Enter 1 for k into the formula to find the second solution.

f) Try another. Find the cube root of $z = 27(\cos 120° + i \sin 120°)$. (Since this problem is in degrees, change $2\pi k$ into $360k$.)

g) If we take a cube root, there will be three complex roots. To find them, we will start with $k = 0$. Enter 0 for k into the formula to find the first solution.

h) The second solution will occur when $k = 1$. Enter 1 for k into the formula to find the second solution.

i) Finally, the third solution will occur when $k = 2$. Enter 2 for k into the formula to find the third solution.

j) Try one more. Find the fourth root of $z = 16(\cos 80° + i \sin 80°)$. (Since this problem is in degrees, change $2\pi k$ into $360k$.)

k) If we take a fourth root, there will be four complex roots. To find them, we will start with $k = 0$. Enter 0 for k into the formula to find the first solution.

l) The second solution will occur when $k = 1$. Enter 1 for k into the formula to find the second solution.

m) The third solution will occur when $k = 2$. Enter 2 for k into the formula to find the third solution.

n) Finally, the third solution will occur when $k = 3$. Enter 3 for k into the formula to find the fourth solution.

As I'm sure you've realized, for an nth root, the values of k will go up to $n - 1$.

Trig

Active Learning: 10.8a

We begin a new section on vectors. Vectors are simply line segments, but with one addition. The direction of these line segments matters too. Here's a vector.

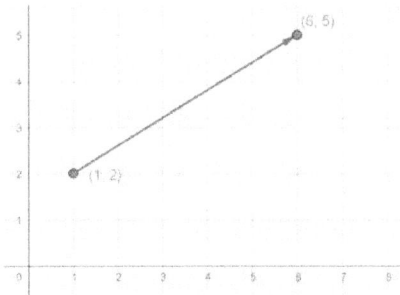

The first thing we want to do with vectors is to put them in standard position—called a position vector. Position vectors start at the origin and can be written in a format which we can be more easily understand.

Moving a vector to the origin is easy. Just subtract the two x coordinates and the two y coordinates.

$$v = \langle x_2 - x_1, y_2 - y_1 \rangle$$

So here is the vector above moved to the origin.

$$v = \langle 6 - 1, 5 - 2 \rangle$$

$$v = \langle 5, 3 \rangle$$

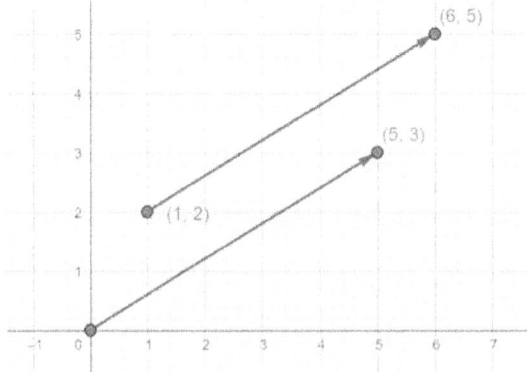

a) Move the following vectors to the origin:

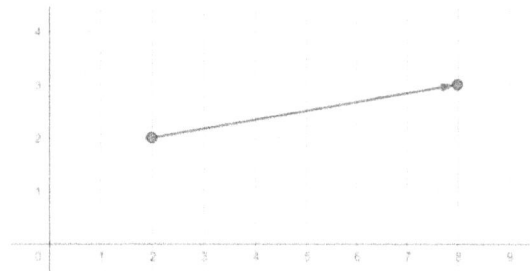

Position Vector:

$$v = \langle \underline{}, \underline{} \rangle$$

b) Be careful on this next one. The ending point must always be (x_2, y_2).

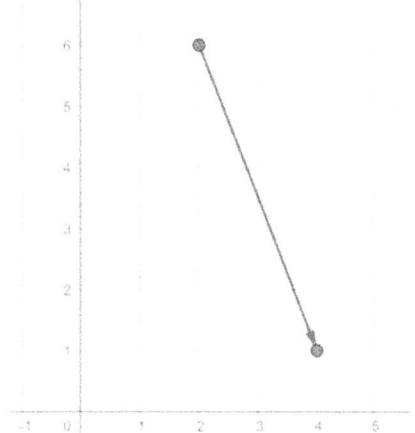

Position Vector:

$$v = \langle \underline{}, \underline{} \rangle$$

c) Two vectors are equal if their position vectors are identical. Determine if the following vectors are the same. Vector u starts at P_1 and ends at P_2. Vector v starts at P_3 and ends at P_4.

$$P_1 = (-1, -2), P_2 = (6, 7); \ P_3 = (3, 4), P_4 = (10, 13)$$

Below are two vectors. u is the larger and v is the smaller.

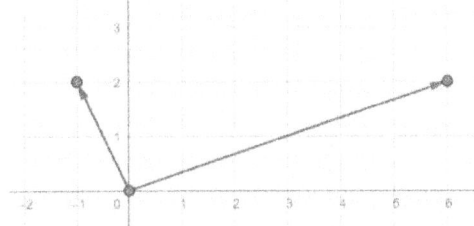

Technically, to add them $(u + v)$, we place the second vector at the end of the first.

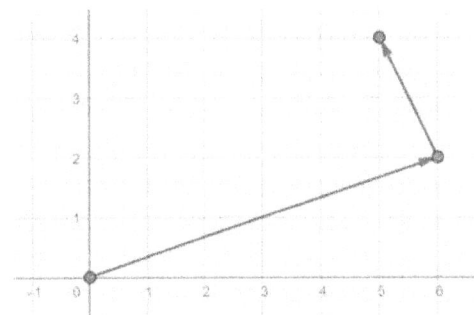

The new vector formed is now their sum.

d) What is the value of $u + v$?

e) Show that you get the same thing by simply adding the x-portion of the vectors and the y-portion of the vectors.

$u = \langle 6, 2 \rangle$ and $v = \langle -1, 2 \rangle$

f) Add the following vectors by simply adding their components:

$u = \langle -2, 7 \rangle$ and $v = \langle 6, 12 \rangle$

Next, we will look at subtracting vectors. To subtract vectors $u - v$, we change the signs of vector v.

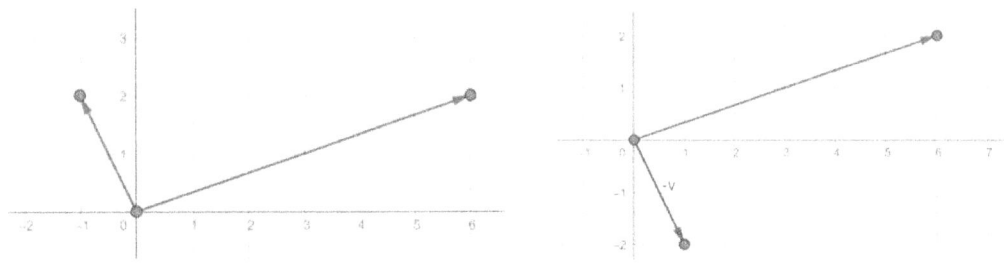

And then put it at the end of the first vector. The new vector is $u - v$.

g) 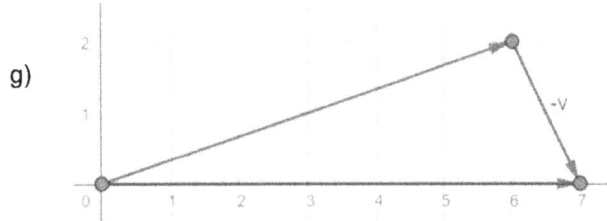 What is the value of $u - v$?

h) Show that you get the same thing by simply subtracting the x-portion of the vectors and the y-portion of the vectors.

$u = \langle 6, 2 \rangle$ and $v = \langle -1, 2 \rangle$

Subtract the following vectors by subtracting their components:

$u = \langle -2, 7 \rangle$ and $v = \langle 6, 12 \rangle$

i) $u - v =$

j) $v - u =$

Finally, we want to look at the magnitude and direction of a vector. The idea is identical to what we've seen before. The magnitude is the same as the modulus of a complex number. The direction is the angle which the vector forms from the positive x-axis.

k)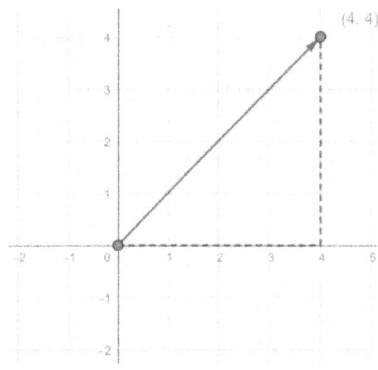

If the vector is a position vector (starting at the origin), we can form a right triangle. We can then use the Pythagorean Theorem to find the magnitude (the hypotenuse) of the vector. Here is the notation.

$$|v| = \sqrt{x^2 + y^2}$$

Find the magnitude of the vector in the diagram.

$|v| =$

To find the direction of the vector, we need to find the angle θ which is formed with the positive x-axis. The trick we learned earlier was to use tangent.

$$\tan \theta = \frac{y}{x}$$

$$\tan \theta = \frac{4}{4}$$

$$\tan \theta = 1$$

l) Find the angle θ which has a tangent ratio of 1. From the original vector, you know that it must be in Quadrant I.

m) Try one on your own. Find the magnitude and direction of the vector, v, from the diagram.

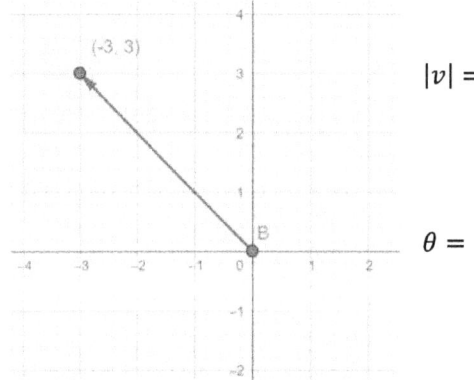

$|v| =$

$\theta =$

In this variation of the problem, the vector is not a position vector. Instead, you are given the initial and terminal points. To find the magnitude, the formula essentially changes to the distance formula. Remember, that the point (x_2, y_2) must be the terminal point of the vector.

$$|v| = \sqrt{(x_2 - x_1)^2 + (y_2 - y_1)^2}$$

n) Find the magnitude and direction of the following vector with initial point $(4, 6)$ and terminal point $(-1, 1)$.

$|v| =$

$\theta =$

Here is one last variation. You are asked to find the magnitude and direction of vector; however, the value of tangent is not on the unit circle. Instead, you will need to use inverse tangent on your calculator.

o) Give the magnitude and direction of the following position vector: $\langle 3, -2 \rangle$. (Find the angle in degrees.)

$|v| =$

$\theta =$

Trig

Active Learning: 10.8b

In the last activity, we learned how to add and subtract vectors. We can also multiply them by a scalar.

$$5\langle -2, 4\rangle$$

This simply means that we will multiply both the x portion and the y portion of the vector by that number.

$$5\langle -2, 4\rangle = \langle 5(-2), 5(4)\rangle = \langle -10, 20\rangle$$

Multiply the following vectors by the scalar.

a) $3\langle -6, -2\rangle =$

b) $6\langle 7, -3\rangle =$

We may also be asked to combine scalar multiplication and addition/subtraction. It follows the typical order. Multiply the scalars first and then do the addition or subtraction.

$$u = \langle 4, -2\rangle \qquad v = \langle 3, 5\rangle$$

Find the following:

c) $2u + 3v =$

d) $4u - 2v =$

Vectors can also be written in a component form. For instance, the vector u from the last problem can be written as follows:

$$u = \langle 4, -2\rangle = 4i - 2j$$

Where the x portion of the vector is in front of i, and the y portion of the vector is in front of j. Write the vector v in component form.

e) $v = \langle 3, 5\rangle =$

Given the initial and terminal points of a vector, write it in component form.

Initial = $(-2, 5)$, Terminal = $(-6, 9)$

f) $v =$

Finally, we want to look at the idea of a unit vector. A unit vector is in the same direction as a vector but has a magnitude of 1.

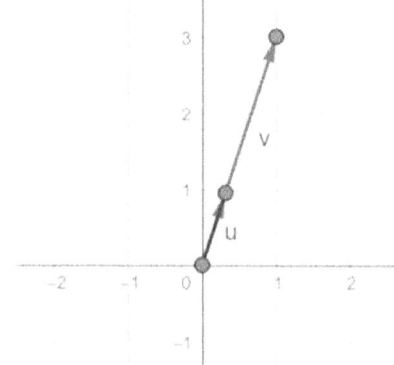

In this graph, the vector v has a unit vector u. To find a unit vector, divide each of the components by the magnitude of the vector.

$$u = \frac{v}{|v|}$$

g)   Using the vector v from the graph, what is its magnitude.

$|v| =$

Now, divide each component of vector v by that magnitude to make the unit vector.

h) $u =$ ___$i +$ ___j

Now, find the magnitude of u to see that it has a magnitude of 1.

i) $|u| =$

Find a unit vector for each of the following vectors:

j) $\langle 4, -2 \rangle$ k) $\langle 3, 5 \rangle$

$u =$ $u =$

Trig

Active Learning: 10.8c

In our last activity, we saw how we can turn a vector, $v = \langle 3, 4 \rangle$, into a component vector, $v = 3i + 4j$.

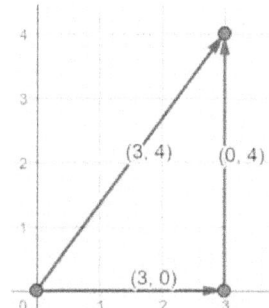

But, as we've done several times now, we could use trig functions to explain each of the components of the vector.

$$\cos \theta = \frac{x}{|v|}$$

$$\sin \theta = \frac{y}{|v|}$$

Solve these equations for x and y.

a) $x =$

b) $y =$

We can then write a component vector if we are given the magnitude and direction of the vector. Use your equations for x and y to find the component vector.

c) A vector has length (magnitude) of 5 and a direction of 150°. Give the component vector.

$v = $ ___ $i + $ ___ j

Below is the equation for a dot product of two vectors: $v = \langle a_1, b_1 \rangle$ and $w = \langle a_2, b_2 \rangle$.

$$v \cdot w = a_1 a_2 + b_1 b_2$$

A dot product turns a vector into a scalar (a number with no direction.) It is a not a very intuitive idea but in the following exercise I'll try to show you what it means.

171

Under each vector, write what quadrant (I, II, III, or IV) it would be in on a graph. Then, find the dot product of the following vectors:

d) $v = \langle 2, 5 \rangle$ and $w = \langle 2, 7 \rangle$

e) $v = \langle 2, 5 \rangle$ and $w = \langle -2, 7 \rangle$

f) $v = \langle 2, 5 \rangle$ and $w = \langle -2, -7 \rangle$

g)   Of the three dot products which you have created, which set created the largest number?

h)   Look at the quadrants of the pair you have chosen, what do they have in common?

Although it may not seem like it, dot products actually give information about the angle between the two vectors. If the angle is smaller, you get a larger value for the dot product. Something interesting also happens in the case below. Find the dot product of the following:

i) $v = \langle 5, 0 \rangle$ and $w = \langle 0, 7 \rangle$

j)   What is the angle between these two vectors?

When you get such a dot product, you always have that angle between the vectors. In those cases, the vectors are called orthogonal.

The idea that dot products are related to the angle between vectors can be seen more clearly in the formula below.

$$\cos\theta = \frac{v \cdot w}{|v| \cdot |w|}$$

Here θ is the measure of the angle between the two vectors. Use the formula to find the measure of the angle between the two vectors. The denominator of the formula involves the magnitudes of the vectors. (Hint: Solve the right side of the equation and then take the inverse cosine to find the angle.)

k) $v = \langle 2, 5 \rangle$ and $w = \langle 2, 7 \rangle$

l) $v = \langle 2, 5 \rangle$ and $w = \langle -2, 7 \rangle$

m) $v = \langle 5, 0 \rangle$ and $w = \langle 0, 7 \rangle$ (I know you already know this one, but show your work to see that you get the same result.)

Finally, determine if the following component vectors are orthogonal. (You don't need to use the cosine formula; their simple dot product will tell you.)

n) $v = 4i + 3j$ and $w = 6i - 10j$

o) $v = -2i + 1j$ and $w = 11i + 22j$

Trigonometry

Active Learning Answer Key

7.1a

a) 2, 2; 3, 3; 4, 4

b) b-The radius of the circle

c) 2.5 radians

d) 81967.2 feet

e) $\frac{7\pi}{6}$ or 3.67 radians

f) 15°

7.1b

a) They are all the same angle. 390° is one additional trip around the circle. 750° is a second additional trip around the circle.

b) 140°

c) 300°

d) 2π

e) $\frac{2\pi}{1}$

f) $\frac{6\pi}{3}$

g) $\frac{12\pi}{6}$

h) $\frac{2\pi}{3}$

i) $\frac{7\pi}{6}$

7.1c

a) 3 radians

b) $s = r\theta$

c) $\frac{3\pi}{4}$

d) $\frac{8\pi}{3}$

e) $\frac{7\pi}{6}$

f) $\frac{14\pi}{3}$

g) $A = \pi r^2$

h) $\frac{60}{360} = \frac{1}{6}$

i) $\frac{8\pi}{3}$

j) $\frac{1}{12}$

k) $\frac{4\pi}{3}$

l) $\frac{\pi}{120}$ $radians/second$ or $.026\ radians/second$

m) The same.

n) The runner on the outside of the circle.

o) The runner on the outside of the circle.

p) 50 $meters/minute$

q) $\frac{1}{5}$ $radians/minute$

r) 6π

s) 200π

t) 600π $radians/minute$

u) 4480 $inches/minute$

7.2a

a) $\frac{4.47}{4} = \frac{x}{10}$

b) $x = 11.175$

c) .4474

d) .8949

e) .5

f) .4474

g) .8989

h) .5

i) The sine ratios are the same; the cosine ratios are the same; the tangent ratios are the same.

j) $\frac{3}{5} = .6$

k) $\frac{4}{5} = .8$

l) $\frac{3}{4} = .75$

m) $\frac{5}{3} = 1.67$

n) $\frac{5}{4} = 1.25$

o) $\frac{4}{3} = 1.33$

p) $\frac{5}{13} = .3846$

q) $\frac{13}{5} = 2.6$

r) $\frac{12}{13} = .9231$

s) $\frac{13}{12} = 1.083$

t) $\frac{5}{12} = .4167$

u) $\frac{12}{5} = 2.4$

7.2b

a) $\frac{\sqrt{2}}{2}$

b) $\frac{\sqrt{2}}{2}$

c) 1

d) $\frac{1}{2}$

e) $\frac{\sqrt{3}}{2}$

f) $\frac{1}{\sqrt{3}} = \frac{\sqrt{3}}{3}$

g) $\frac{\sqrt{3}}{2}$

h) $\frac{1}{2}$

i) $\sqrt{3}$

j) They are the same.

k) They are both referring to the same sides.

l) They are the same.

m) They are both referring to the same sides.

n) $\frac{\sqrt{3}}{3}$

o) They are the same.

p) $\sqrt{3}$; $\sqrt{3}$

q) 2; 2

r) They are the same.

s) $\frac{2}{\sqrt{3}} = \frac{2\sqrt{3}}{3}$; $\frac{2}{\sqrt{3}} = \frac{2\sqrt{3}}{3}$

t) They are the same.

u) They add to 90°

v) b is not true. The angles should add to $\frac{\pi}{2}$.

w) $\sec 60 = 3$

x) $\cos \frac{\pi}{6} = \frac{\sqrt{3}}{2}$

y) $\cot 90 = 0$

7.2c

a) b

b) 29.79 feet

c) a

d) 74.6 feet

e) elevation; depression

f) 12.28 feet

g) elevation

h) 22 feet

i) depression

7.3a

a) 5

b) The distance formula is the Pythagorean Theorem.

c) $\frac{4}{5}$

d) $\frac{3}{5}$

e) $\frac{4}{3}$

f) 13

g) $\frac{5}{13}$

h) $\frac{12}{13}$

i) $\frac{5}{12}$

j) $\sin a = y$

k) $\cos a = x$

l) $x = \frac{1}{2}, y = \frac{\sqrt{3}}{2}$

m) $\cos 60 = \frac{1}{2}; \sin 60 = \frac{\sqrt{3}}{2}$

n) $x = \frac{\sqrt{3}}{2}; y = \frac{1}{2}$

o) $\cos 30 = \frac{\sqrt{3}}{2}; \sin 30 = \frac{1}{2}$

p) $x = \frac{\sqrt{2}}{2}; y = \frac{\sqrt{2}}{2}$

q) $\cos 45 = \frac{\sqrt{2}}{2}; \sin 45 = \frac{\sqrt{2}}{2}$

r) $(-1, 0)$

s) $x = -1$

t) $y = 0$

u) $\frac{0}{-1} = 0$

v) $(0, -1)$

w) $x = 0$

x) $y = -1$

y) $\frac{-1}{0} = undefined$

z) It was undefined because it created a negative in the denominator.

aa) $(-1, 0)$

bb) $x = -1$

cc) $y = 0$

dd) $\frac{0}{-1} = 0$

7.3b

a) positive

b) positive

c) negative

d) negative

e) positive

f) negative

g) negative

h) positive

i) $x^2 + y^2 = r^2$

j) $(\cos \alpha)^2 + (\sin \alpha)^2 = (1)^2$ or $\cos^2 \alpha + \sin^2 \alpha = 1$

k) $\cos^2 \alpha = \frac{16}{25}$

l) $\pm \frac{4}{5}$

m) negative

n) $-\frac{4}{5}$

o) $-\frac{12}{13}$

n)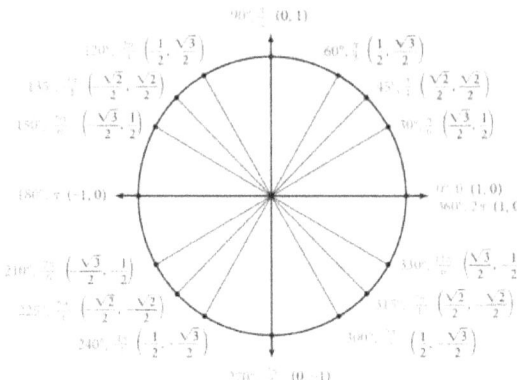

7.3c

a) 45°

b) 65°

c) $\frac{\pi}{12}$

d) $\frac{\pi}{6}$

e) 60°

f) $-\frac{1}{2}$

g) $\frac{\sqrt{3}}{2}$

h) negative

i) positive

j) 45°

k) $-\frac{\sqrt{2}}{2}$

l) $-\frac{\sqrt{2}}{2}$

m) Both x and y are negative in quadrant III.

7.4a

a) 1

b) $-\sqrt{3}$

c) $-\frac{2}{\sqrt{2}} = -\frac{2\sqrt{2}}{2} - \sqrt{2}$

d) $-\frac{2}{\sqrt{2}} = -\frac{2\sqrt{2}}{2} - \sqrt{2}$

e) 1

f) $\frac{2\sqrt{3}}{3}$

g) -2

h) $-\frac{1}{\sqrt{3}} = -\frac{\sqrt{3}}{3}$

7.4b

a) even

b) a

c) odd

d) b

e) $\csc x$

f) $\sec x$

g) $\csc(-x) = -\csc(x)$

h) $\sec(-x) = \sec(x)$

i) b

j) odd

k) b

l) odd

m) $\sec(-x) = 2$

n) $\cot(-x) = -5$

7.4c

a) $1 + tan^2(x) = sec^2(x)$

b) $cot^2(x) + 1 = csc^2(x)$

c) 12

d) negative

e) $-\frac{12}{13}$

f) $-\frac{13}{12}$

g) $\frac{13}{5}$

h) $-\frac{12}{5}$

i) $-\frac{5}{12}$

j) The missing side is -4.

k) $-\frac{4}{5}$

l) $-\frac{3}{5}$

m) $\frac{4}{3}$

n) $-\frac{5}{4}$

u) $-\frac{5}{3}$

p) $\frac{3}{4}$

8.1a

a) The first two are functions (the square function and the absolute value). The circle is not a function.

b) The x-axis

c) 1

d) 2π

e) The x-axis, with is the line $y = 0$.

f) 1

g) 2π

h) It has been vertically stretched by 2.

i) Amplitude

j) Flipped upside down and vertically compressed by $\frac{1}{2}$.

k) Horizontally stretched by .5 (by a factor of 2).

l) Period

m) Horizontally compressed by 2 (by a factor of ½).

n) Period

o) Shifted up 1.

p) The midline is now $y = 1$.

q) Shifted down 1.

r) The midline is now $y = -1$.

s) Shifted right by $\frac{\pi}{3}$.

t) Shifted left by $\frac{\pi}{3}$.

u) An amplitude of 3.

v) A phase shift to the right $\frac{\pi}{2}$.

w) Flipped over the x axis.

x) A horizontal compression resulting in a period of π.

y) A vertical shift up 3. The midline is now $y = 3$.

z) Period: π

aa) Period: $\frac{2\pi}{3}$

bb) Period: $\frac{2\pi}{\frac{1}{2}} = 4\pi$

8.1b

a) .5

b) Right $\frac{3\pi}{2}$

c) 2π

d) $y = 1$

e) No

f) 1

g) Left $\frac{\pi}{3}$

h) 4π

i) $y = 0$

j) No

k) 10

l) Left $\frac{7\pi}{6}$

m) $\frac{\pi}{3}$

n) $y = 0$

o) Yes

p) Yes, one has a Greatest Common Factor pulled out.

q) Right 1

r) Left 2π

s) Left 4

t)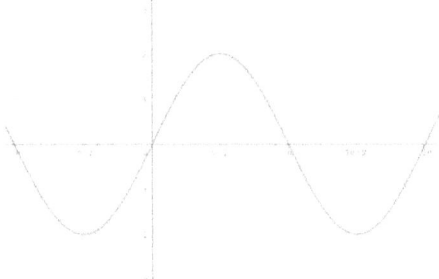

u)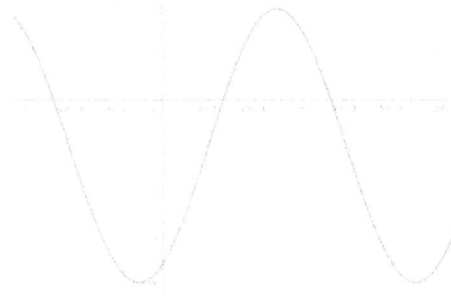

v) $B = 3$

w) $B = 4$

x) $\frac{2\pi}{3}$

y) $B = 3$

z) $y = 1$

aa) 2

bb) Sine

cc) $f(x) = 2\sin(3x) + 1$

8.2a

a) π

b) $\cos(x)$ is zero at those values which would make tangent undefined.

c) The values of y extend on to infinity, so there can be no amplitude.

d) Stretch factor is 2

e) Period: $\frac{\pi}{3}$

f) Phase Shift: $\frac{2\pi}{3}$ to the right

g) Midline: $y = -3$

h) Stretch factor is 5

i) Period: $\frac{\pi}{2}$

j) Phase Shift: $\frac{3\pi}{2}$ to the right

k) Midline: $y = 4$

l) $(0, 0)$

m) $\left(\frac{\pi}{4}, 1\right)$

n) $\left(-\frac{\pi}{4}, -1\right)$

o) $(0, -3)$

p) $\left(\frac{\pi}{12}, -1\right)$

q) $\left(-\frac{\pi}{12}, -5\right)$

r) $x = \frac{\pi}{6}$

s) $x = -\frac{\pi}{6}$

t)

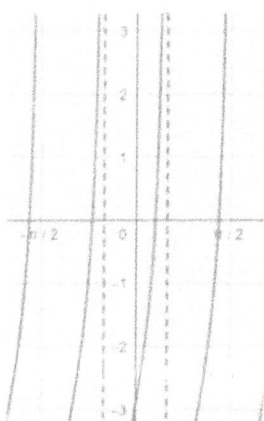

8.2b

a) Since $\sec(x) = \frac{1}{\cos(x)}$, secant has asymptotes wherever cosine is equal to zero.

b)

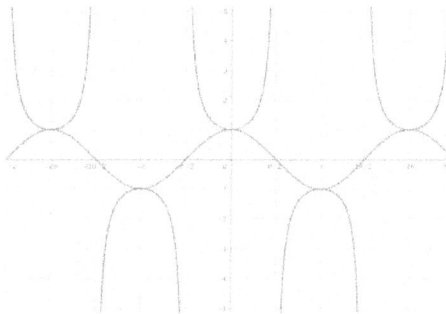

c) Stretch: 2

d) Period: π

e) Phase Shift: $\frac{\pi}{2}$ to the right

f) Midline: $y = 0$

g)

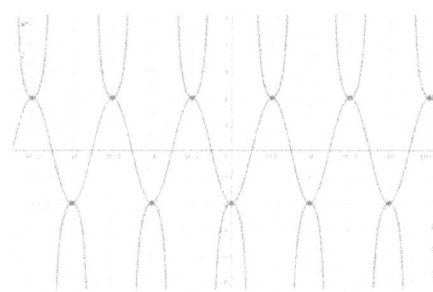

h)

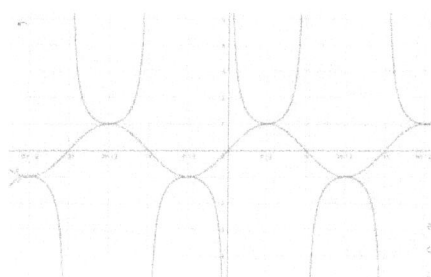

i) Stretch: 2

j) Period: $\frac{2\pi}{3}$

k) Phase Shift: $\frac{\pi}{4}$ to the left

l) Midline: $y = 0$

m)

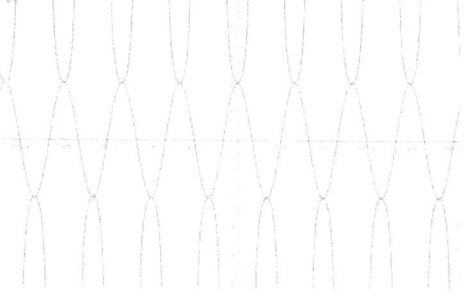

8.3a

a) The following would have an inverse.

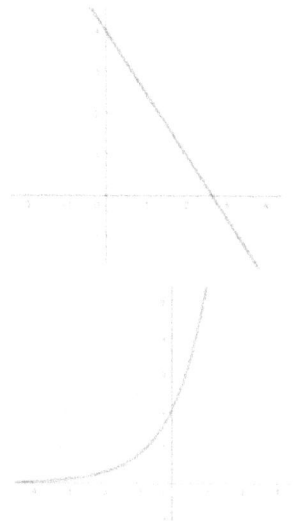

b) sine function. Not one to one.

c) cosine function. Not one to one.

d) Domain: $[-\frac{\pi}{2}, \frac{\pi}{2}]$

e) Range: $[-1, 1]$

f) Domain: $[0, \pi]$

g) Range: $[-1, 1]$

h) Domain: $[-1, 1]$

i) Range: $[-\frac{\pi}{2}, \frac{\pi}{2}]$

j) Domain: $[-1, 1]$

k) Range: $[0, \pi]$

l) $\frac{\pi}{6}$ and $\frac{5\pi}{6}$

m) $\frac{\pi}{6}$

n) $\frac{2\pi}{3}$ and $\frac{4\pi}{3}$

o) $\frac{2\pi}{3}$

p) 2.214

q) .1911

r) 1.3258

s) .8143

t) .8478

8.3b

a) $-\frac{1}{2}$

b) $\frac{2\pi}{3}$ and $\frac{4\pi}{3}$

c) $\frac{2\pi}{3}$

d) $-\frac{\sqrt{3}}{2}$

e) $\frac{4\pi}{3}$ and $\frac{5\pi}{3}$

f) $\frac{5\pi}{3}$

g) b

h) The missing side is 4

i) $\frac{4}{5}$

j) The missing side is 13

k) $\frac{5}{13}$

8.3c

a) $\sqrt{2x+1}$

b) $\frac{\sqrt{2x+1}}{x+1}$

c) $\sqrt{1-x^2}$

d) $\frac{\sqrt{1-x^2}}{1} = \sqrt{1-x^2}$

e) 1

f) $\frac{1}{\sqrt{x^2+1}}$

9.1a

a) $y = -2x$

b) $3x + (-2x) = 1$

c) $y = -2; (1, -2)$

d)
$\cos x \dfrac{\sin x}{\cos x} = \sin x$

e)
$\dfrac{\frac{\cos x}{\sin x}}{\frac{1}{\sin x}} = \dfrac{\cos x}{\sin x} \cdot \dfrac{\sin x}{1} = \cos x$

f)
$\dfrac{\cos^2 x}{\cos^2 x} + \dfrac{\sin^2 x}{\cos^2 x} = \dfrac{\cos^2 x + \sin^2 x}{\cos^2 x}$

g)
$\dfrac{\cos^2 x + \sin^2 x}{\cos^2 x} = \dfrac{1}{\cos^2 x} = \sec^2 x$

h) $\dfrac{1-\sin^2 x}{\cos} = \dfrac{\cos^2 x}{\cos} = \cos x$

i)
$\dfrac{\frac{\sin x}{\cos x}}{\frac{\cos x}{\sin x} + \frac{\sin x}{\cos x}} = \dfrac{\frac{\sin x}{\cos x}}{\frac{\cos^2 x}{\sin x \cos x} + \frac{\sin^2 x}{\sin x \cos x}}$

$= \dfrac{\frac{\sin x}{\cos x}}{\frac{1}{\sin x \cos x}}$

$= \dfrac{\sin x}{\cos x} \cdot \dfrac{\sin x \cos x}{1} = \sin^2 x$

9.1b

a) a

b) b

c) $-\csc x$

d) $\sec x$

e) $-\tan x$

f) $-\cot x$

g) $--\sin x \cot x \cos x = \sin x \dfrac{\cos x}{\sin x} \cos x = \cos^2 x$

h) $(x-1)(x+1)$

i) $(\tan x - 1)(\tan x + 1)$

j) $\dfrac{(\tan x - 1)(\tan x + 1)}{\sin x (\tan x + 1)} = \dfrac{\tan x - 1}{\sin x} = \dfrac{\frac{\sin}{\cos} - \frac{\cos}{\cos}}{\sin} =$

$\dfrac{\frac{\sin - \cos}{\cos}}{\sin} = \dfrac{\sin - \cos}{\cos} \cdot \dfrac{1}{\sin} = \dfrac{\sin - \cos}{\sin \cos} = \dfrac{\sin}{\sin \cos} - \dfrac{\cos}{\sin \cos}$

$\dfrac{1}{\cos} - \dfrac{1}{\sin} = \sec - \csc$

k) $1 - \sin^2 x$

l) $1 - \sin^2 x = \cos^2 x$

m) $\dfrac{\cos}{1+\sin} \cdot \dfrac{1-\sin}{1-\sin} = \dfrac{\cos(1-\sin)}{1-\sin^2} = \dfrac{\cos(1-\sin)}{\cos^2} =$

$\dfrac{1-\sin}{\cos}$

183

9.2a

a) $30°, 45°, 60°$ or $\frac{\pi}{6}, \frac{\pi}{4}, \frac{\pi}{3}$

b) $u = 60, v = 45$

c) $u = \frac{\pi}{4}, v = \frac{\pi}{6}$

d) $u = 135, v = 120$

e) $u = 135, v = 30$

f) $u = \frac{\pi}{4}, v = \frac{\pi}{6}$

g) $u = 225, v = 60$

It is possible to get multiple equivalent answers for sum and difference problems. If you are unsure if your answer is the same, use your calculator to determine the values.

h) $\frac{\sqrt{6}-\sqrt{2}}{4}$

i) $\frac{\sqrt{2}+\sqrt{6}}{4}$

j) $\frac{-\sqrt{2}-\sqrt{6}}{4}$

k) $\frac{-\sqrt{6}+\sqrt{2}}{4}$

l)
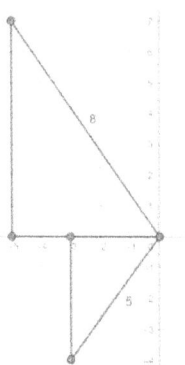

m) $\cos u = -\frac{3}{5}$

n) $\sin v = \frac{\sqrt{39}}{8}$

o) $\cos(u - v) = \frac{15 + 4\sqrt{39}}{40}$

9.2b

a) a

b) $\frac{1}{\sqrt{3}}$

c) 1

d) $\frac{1+\sqrt{3}}{\sqrt{3}-1}$

e) $\sin x$

f) $\csc x$

g) $\tan x$

h) $\sin 18$

i) $\csc \frac{5\pi}{14}$

j) $\tan \frac{7\pi}{18}$

k) $\frac{\sin\alpha\cos\beta - \cos\alpha\sin\beta}{\cos\alpha\cos\beta} = \frac{\sin\alpha\cos\beta}{\cos\alpha\cos\beta} - \frac{\cos\alpha\sin\beta}{\cos\alpha\cos\beta} = \frac{\sin\alpha}{\cos\alpha} - \frac{\sin\beta}{\cos\beta} = \tan\alpha - \tan\beta$

9.3a

a) $\sqrt{39}$

b) $-\frac{\sqrt{39}}{8}$

c) $\sin\alpha = \frac{5}{8}$

d) $\cos\alpha = -\frac{\sqrt{39}}{8}$

e) $-\frac{5\sqrt{39}}{32}$

f) $\frac{7}{32}$

g) $\sin\alpha = \frac{3}{5}$

h) $\cos\alpha = -\frac{4}{5}$

i) $-\frac{24}{25}$

j) $\frac{7}{25}$

k) $-\frac{24}{7}$

l) $4\sin\frac{\pi}{4}$

m) $3\cos\frac{\pi}{4}$

n) $\frac{\sec\alpha(2\sin\alpha\cos\alpha)}{2} = \frac{\frac{1}{\cos\alpha}(2\sin\alpha\cos\alpha)}{2} = \sin\alpha$

9.3b

a) $\frac{1+\cos(8x)}{2}$

b) $\frac{1-2\cos(12x)+\cos^2(12x)}{4}$

c) $\frac{1-2\cos(12x)+\cos^2(12x)}{4} = \frac{1}{4} - \frac{1}{2}\cos(12x) + \frac{1}{4}\left(\frac{1+\cos(24x)}{2}\right) = \frac{1}{4} - \frac{1}{2}\cos(12x) + \frac{1}{8} + \frac{1}{8}\cos(24x)$

$\frac{3}{8} - \frac{1}{2}\cos(12x) + \frac{1}{8}\cos(24x)$

d)
$\cos^3(4x) = (\cos(4x))(\cos^2(4x))$
$= (\cos(4x))\left(\frac{1+\cos(8x)}{2}\right)$
$= \frac{1}{2}(\cos(4x))(1+\cos(8x))$

9.3c

a) 3

b) $\frac{3}{5}$

c) $\pm\sqrt{\frac{1}{5}}$

d) Quadrant II

e) $+\sqrt{\frac{1}{5}}$

f) $-\sqrt{\frac{4}{5}}$

g) $\frac{\sqrt{3}}{2}$

h) $+\sqrt{\frac{2+\sqrt{3}}{2}}$

i) $+\sqrt{\frac{2+\sqrt{2}}{4}}$

j) $+\sqrt{\frac{2+\sqrt{3}}{4}}$

k) $-\sqrt{\frac{2-\sqrt{2}}{2+\sqrt{2}}}$

9.4a

a)
$2\cos\alpha\cos\beta = \cos(\alpha-\beta) + \cos(\alpha+\beta)$
$\cos\alpha\cos\beta = \frac{1}{2}[\cos(\alpha-\beta) + \cos(\alpha+\beta)]$

b) $\frac{1}{2}\left[\cos\left(\frac{3x}{3}\right) + \cos\left(\frac{13x}{3}\right)\right]$

c) $2[\sin(8\theta) + \sin(4\theta)]$

d) $4[\sin(9\theta) - \sin(-5\theta)] = 4[\sin(9\theta) + \sin(5\theta)]$ (I used the odd identity for sine to get rid of the negative angle.)

e) $\frac{1}{4}\left[\cos\left(\frac{12\pi}{12}\right) + \cos\left(\frac{14\pi}{12}\right)\right] = \frac{1}{4}\left[\cos(\pi) + \cos\left(\frac{7\pi}{6}\right)\right] = \frac{1}{4}\left[-1 - \frac{\sqrt{3}}{2}\right] = \frac{-2-\sqrt{3}}{8}$

9.4b

a) $2\sin(3\theta)\cos(5\theta)$

b) $2\cos(5\theta)\cos(2\theta)$

c) $2\sin(45°)\sin(-30°) = \frac{\sqrt{6}}{2}$

9.5a

a) $\frac{1}{2}$

b) $\frac{1}{2}$

c) $\frac{1}{2}$

d) $\frac{1}{2}$

e) You are getting the same answer each time because of coterminal angles.

f) 1

g) 1

h) 1

i) 1

j) The period of tangent is π, so the coterminal angles repeat every half of a circle.

k) We need to include any coterminal angles which would also have the same value.

l) $\frac{\pi}{3}$ and $\frac{2\pi}{3}$

m) Since we are working with tangent the coterminal angles are now every half of a circle.

n) $x = \frac{\sqrt{2}}{2}$

o) $\frac{\pi}{4} \pm 2\pi k$ and $\frac{3\pi}{4} \pm 2\pi k$

p) π

q) $\frac{7\pi}{6}$ and $\frac{11\pi}{6}$

r) .1002

s) 3.041

t) 1.369

u) 4.914

9.5b

a) $\frac{\pi}{12}$

b) $\frac{25\pi}{12}$

c) $\frac{11\pi}{12}$

d) $\frac{35\pi}{12}$

e) yes

f) yes

g) $y = 2x + 4$

h) $\frac{13\pi}{12}$

i) $\frac{11\pi}{12} + \pi k$

j) $\frac{23\pi}{12}$

k) $\frac{\pi}{6} + \pi k$ and $\frac{5\pi}{6} + \pi k$. So, the unique solutions are: $\frac{\pi}{6}, \frac{5\pi}{6}, \frac{7\pi}{6}, \frac{11\pi}{6}$.

l) $\frac{\pi}{9} + \frac{2\pi}{3}k$ and $\frac{2\pi}{9} + \frac{2\pi}{3}k$. So, the unique solutions are: $\frac{\pi}{9}, \frac{2\pi}{9}, \frac{7\pi}{9}, \frac{8\pi}{9}, \frac{13\pi}{9}, \frac{14\pi}{9}$.

9.5c

a) $x = 0$ and $x = -\frac{1}{2}$

b) $0 + 2\pi k, \pi + 2\pi k, \frac{7\pi}{6} + 2\pi k, \frac{11\pi}{6} + 2\pi k$

c) .6155, -.6155

d) 2.526, 3.757

e) $x = \frac{3}{2}$ and $x = 1$

f) $\sin\theta = \frac{3}{2}$ and $\sin\theta = 1$

g) $\theta = \frac{\pi}{2}$

h) $x = 0$ and $x = \frac{1}{2}$

i) $\theta = 0, \pi, \frac{\pi}{6}, \frac{5\pi}{6}$

j) .8861, -.8861, 2.2555, 4.0277

10.1a

a) 17.36

b) 17.36

c) 17.36

d) They are the same.

e) The angle is paired with the side opposite from it.

f) .0576

g) .0576

h) .0576

i) They are the same.

j)

$$\frac{\sin\alpha}{a} = \frac{\sin\beta}{b} = \frac{\sin\gamma}{c}$$

$$\frac{a}{\sin\alpha} = \frac{b}{\sin\beta} = \frac{c}{\sin\gamma}$$

k) 39°

l) $a = 23.85$

m) $b = 34.62$

n) $b = 38.12$

o) $c = 29.28$

10.1b

a) $\sin\beta = 1.915$

b) The range of sine is $[-1, 1]$, so this is impossible.

c) 83.2°

d) 16.8°

e) $c = 35.2$

f) 96.784°

g) $\gamma' = 3.216°$

h) $c' = 6.84$

i) $\alpha = 14.33$

j) $\alpha' = 165.67$

k) It exceeds 180° and there is no room left for a third angle.

l) $\sin\alpha = \frac{h}{c}$

m) $h = c \cdot \sin\alpha$

n) $A = \frac{1}{2} b \cdot h$

o) $h = c \cdot \sin\alpha$

p) $A = \frac{1}{2} b \cdot c \cdot \sin\alpha$

q) 2288.87

10.2a

a) We have the angle of the missing side.

b) $a = 22.7$

c) $b = 47.8°$

d) $C = 85.2°$

e) $\cos A = \frac{a^2 - b^2 - c^2}{-2bc}$

f) $\cos B = \frac{b^2 - a^2 - c^2}{-2ac}$

g) $\cos C = \frac{c^2-a^2-b^2}{-2ab}$

h) $A = 47°$

i) $B = 59°$

j) $C = 74°$

k) Quadrants I and IV

l) Quadrants I and II

10.2b

a) $s = 32.5$

b) $Area = 192.64$

c) $s = 43.4$

d) $Area = 336.64$

10.3a

a)

b)

c) $\cos\theta = \frac{x}{r}$

d) $\sin\theta = \frac{y}{r}$

e) $x = r\cos\theta$

f) $y = r\sin\theta$

g) Pythagorean Theorem

h) $x^2 + y^2 = r^2$

i) $\left(-\frac{3\sqrt{3}}{2}, \frac{3}{2}\right)$

j) $(2\sqrt{2}, -2\sqrt{2})$

k) $r = 2\sqrt{2}$

l) $\theta = \frac{3\pi}{4}$

m) $\left(2\sqrt{2}, \frac{3\pi}{4}\right)$

n) $\left(4\sqrt{2}, \frac{5\pi}{4}\right)$

o) $\left(4, \frac{4\pi}{3}\right)$

10.3b

a) $r = \pm 11$

b) $r = 3\cos\theta$

c) $r = 11\sin\theta$

d) $r = \frac{3}{\sin\theta - 5\cos\theta}$

e) $r = 4\cos\theta$

10.3c

a) $y = 5$

b) $x = \frac{5}{3}$

c) $r = \frac{3y+5}{2}$

d) $x^2 + y^2 = \left(\frac{3y+5}{2}\right)^2$

e) $x^2 + y^2 = (6-2x)^2$

f) $yx = 2$ or $y = \frac{2}{x}$

10.5a

a)

b)
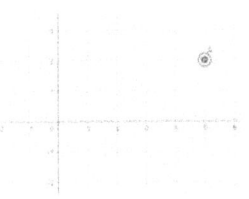

c)
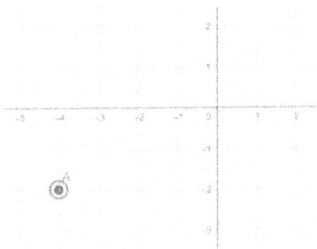

d) $r = \sqrt{13}$

e) $r = \sqrt{41}$

f) $r = 2\sqrt{10}$

g) $\theta = \frac{7\pi}{4}$

h) $\theta = \frac{\pi}{3}$

i) $4\sqrt{2}$

j) $\theta = \frac{9\pi}{4}$

k) $z = 4\sqrt{2}(\cos\frac{3\pi}{4} + i \sin\frac{3\pi}{4})$

l) $r = 2$

m) $\theta = \frac{\pi}{6}$

n) $z = 2(\cos\frac{\pi}{6} + i \sin\frac{\pi}{6})$

o) $z = 6\sqrt{3} + 6i$

p) $z = 2 - 2\sqrt{3}i$

10.5b

a) $r_1 r_2 [\cos(\theta_1)\cos(\theta_2) + \cos(\theta_1)\sin(\theta_2)i + \sin(\theta_1)\cos(\theta_2)i + \sin(\theta_1)\sin(\theta_2)i^2]$

$= r_1 r_2 [(\cos(\theta_1)\cos(\theta_2) - \sin(\theta_1)\sin(\theta_2)) + (\cos(\theta_1)\sin(\theta_2) + \sin(\theta_1)\cos(\theta_2))i]$

$= r_1 r_2 (\cos(\theta_1 + \theta_2) + i \sin(\theta_1 + \theta_2))$

b) $8 cis\left(\frac{11\pi}{6}\right)$ cis is an abbreviation for $\cos\theta + i\sin\theta$.

c) $48\ cis\ 300°$

d) $\frac{1}{2} cis\left(-\frac{5\pi}{6}\right) = \frac{1}{2}\left(\cos\left(\frac{5\pi}{6}\right) - \sin\left(\frac{5\pi}{6}\right)\right)$

e) $\frac{1}{3} cis (-270°) = \frac{1}{3}(\cos(270) - \sin(270))$

f) $r_1 r_1 [\cos(\theta_1)\cos(\theta_1) + \cos(\theta_1)\sin(\theta_1)i + \sin(\theta_1)\cos(\theta_1)i + \sin(\theta_1)\sin(\theta_1)i^2]$

$= r_1 r_1 [(\cos(\theta_1)\cos(\theta_1) - \sin(\theta_1)\sin(\theta_1)) + (\cos(\theta_1)\sin(\theta_1) + \sin(\theta_1)\cos(\theta_1))i]$

$= r^2(\cos(\theta_1 + \theta_1) + i \sin(\theta_1 + \theta_1))$
$= r^2(\cos(2\theta) + i \sin(2\theta))$

g) $64 cis\ (280°)$

h) $4\sqrt{2} cis\left(\frac{35\pi}{4}\right)$

10.5c

a) $z^{\frac{1}{n}} = r^{\frac{1}{n}}[\cos\left(\frac{\theta}{n}\right) + i\sin\left(\frac{\theta}{n}\right)]$

b) $z^{\frac{1}{n}} = r^{\frac{1}{n}}[\cos\left(\frac{\theta}{n} + \frac{2\pi k}{n}\right) + i\sin\left(\frac{\theta}{n} + \frac{2\pi k}{n}\right)]$

c) $z = 6(\cos\left(\frac{7\pi}{12} + \pi k\right) + i\sin\left(\frac{7\pi}{12} + \pi k\right))$

d) $z = 6(\cos\left(\frac{7\pi}{12}\right) + i\sin\left(\frac{7\pi}{12}\right))$

e) $z = 6(\cos\left(\frac{19\pi}{12}\right) + i\sin\left(\frac{19\pi}{12}\right))$

f) $z = 3(\cos(40° + 120°k) + i\sin(40° + 120°k))$

g) $z = 3(\cos(40°) + i\sin(40°))$

h) $z = 3(\cos(160°) + i\sin(160°))$

i) $z = 3(\cos(280°) + i\sin(280°))$

j) $z = 2(\cos(20° + 90°k) + i\sin(20° + 90°k))$

k) $z = 2(\cos(20°) + i\sin(20°))$

l) $z = 2(\cos(110°) + i\sin(110°))$

m) $z = 2(\cos(200°) + i\sin(200°))$

n) $z = 2(\cos(290°) + i\sin(290°))$

10.8a

a) $v = \langle 6, 1 \rangle$

b) $v = \langle 2, -5 \rangle$

c) Both position vectors are $\langle 7, 9 \rangle$, so they are the same.

d) $\langle 5, 4 \rangle$

e) $\langle 5, 4 \rangle$

f) $\langle 4, 19 \rangle$

g) $\langle 7, 0 \rangle$

h) $\langle 7, 0 \rangle$

i) $\langle -8, -5 \rangle$

j) $\langle 8, 5 \rangle$

k) $|v| = 4\sqrt{2}$

l) $\theta = \frac{\pi}{4}$

m) $|v| = 3\sqrt{2}; \theta = \frac{3\pi}{4}$

n) $|v| = 5\sqrt{2}; \theta = \frac{5\pi}{4}$

o) $|v| = \sqrt{13}; \theta = -33.7$

10.8b

a) $\langle -18, -6 \rangle$

b) $\langle 42, -18 \rangle$

c) $\langle 17, 11 \rangle$

d) $\langle 10, -18 \rangle$

e) $v = 3i + 5j$

f) $v = -4i + 4j$

g) $|v| = \sqrt{10}$

h) $u = \frac{1}{\sqrt{10}}i + \frac{3}{\sqrt{10}}j$

i) $|u| = 1$

j) $\langle \frac{4}{2\sqrt{5}}, -\frac{2}{2\sqrt{5}} \rangle = \langle \frac{2}{\sqrt{5}}, -\frac{1}{\sqrt{5}} \rangle$

k) $\langle \frac{3}{\sqrt{34}}, \frac{5}{\sqrt{34}} \rangle$

10.8c

a) $x = |v|\cos\theta$

b) $y = |v|\sin\theta$

c) $v = 5\cos(150)i + 5\sin(150)j$

d) Both are in quadrant I; 39

e) v is in I and u is in II; 31

f) v is in I and u is in III; -39

g) The first set.

h) They are the same quadrant.

i) The dot product is zero.

j) 90°

k) $\theta = 5.86°$

l)) $\theta = 37.7°$

m)) $\theta = 90°$

n) Not orthogonal

o) Orthogonal

www.ingramcontent.com/pod-product-compliance
Lightning Source LLC
Chambersburg PA
CBHW081225170426
43198CB00017B/2714